Geometry of CR-Submanifolds

Aurel Bejancu
Department of Mathematics, Polytechnic Institute of Iași, Romania

Geometry of CR-Submanifolds

D. Reidel Publishing Company
A MEMBER OF THE KLUWER ACADEMIC PUBLISHERS GROUP
Dordrecht / Boston / Lancaster / Tokyo

Library of Congress Cataloging-in-Publication Data

CIP

Bejancu, Aurel, 1946-
Geometry of CR-submanifolds.

(Mathematics and its applications. East European series)
Bibliography: p.
Includes indexes.
1. Submanifolds, CR. 2. Geometry, Differential.
I. Title. II. Series: Mathematics and its applications
(D. Reidel Publishing Company). East European series.
QA649.B44 1986 516.3'6 86-15614
ISBN 90-277-2194-7

Published by D. Reidel Publishing Company
P.O. Box 17, 3300 AA Dordrecht, Holland

Sold and distributed in the U.S.A. and Canada
by Kluwer Academic Publishers,
101 Philip Drive, Assinippi Park, Norwell, MA 02061, U.S.A.

In all other countries sold and distributed
by Kluwer Academic Publishers Group,
P.O. Box 322, 3300 AH Dordrecht, Holland

Printed in The Netherlands

To Ligia, Rodica – Daniela and Aurelian

TABLE OF CONTENTS

EDITOR'S PREFACE

Approach your problems from the right end and begin with the answers. Then one day, perhaps you will find the final question.

'The Hermit Clad in Crane Feathers' in R. van Gulik's *The Chinese Maze Murders.*

It isn't that they can't see the solution. It is that they can't see the problem.

G.K. Chesterton. *The Scandal of Father Brown* 'The point of a Pin'.

Growing specialization and diversification have brought a host of monographs and textbooks on increasingly specialized topics. However, the "tree" of knowledge of mathematics and related fields does not grow only by putting forth new branches. It also happens, quite often in fact, that branches which were thought to be completely disparate are suddenly seen to be related.

Further, the kind and level of sophistication of mathematics applied in various sciences has changed drastically in recent years: measure theory is used (non-trivially) in regional and theoretical economics; algebraic geometry interacts with physics; the Minkowsky lemma, coding theory and the structure of water meet one another in packing and covering theory; quantum fields, crystal defects and mathematical programming profit from homotopy theory; Lie algebras are relevant to filtering; and prediction and electrical engineering can use Stein spaces. And in addition to this there are such new emerging subdisciplines as "experimental mathematics", "CFD", "completely integrable systems", "chaos, synergetics and large-scale order", which are almost impossible to fit into the existing classification schemes. They draw upon widely different sections of mathematics. This programme, Mathematics and Its Applications, is devoted to new emerging (sub)disciplines and to such (new) interrelations as exempla gratia:

- a central concept which plays an important role in several different mathematical and/or scientific specialized areas;
- new applications of the results and ideas from one area of scientific endeavour into another;
- influences which the results, problems and concepts of one field of enquiry have and have had on the development of another.

The Mathematics and Its Applications programme tries to make available a careful selection of books which fit the philosophy outlined above. With such books, which are stimulating rather than definitive, intriguing rather than encyclopaedic, we hope to contribute something towards better communication among the practitioners in diversified fields.

Because of the wealth of scholarly research being undertaken in the Soviet Union, Eastern Europe, and Japan, it was decided to devote special attention to the work emanating from these particular regions. Thus it was decided to start three regional series under the umbrella of the main MIA programme.

The present volume in the MIA (Eastern Europe) series deals with a topic in differential geometry: Cauchy-Riemann submanifolds of Kählerian manifolds and their (many) applications. This is a new field, (the concept was introduced by the author in 1978); it has the vigorousness characteristic of youth, and in spite of its youth it has already manyfold interactions with other parts of mathematics and substantial applications to (pseudo-) conformal mappings and relativity. Also there are interrelations with harmonic maps, deformations of complex structures and more generally the whole field of (real) analysis on complex manifolds. The concept of a CR manifold generalizes both totally real submanifolds and holomorphic submanifolds, both concepts which have proved their worth. However there are not enough of these for many purposes whence the need for the more general notion of CR manifold. This is an up-to-date and self-contained book on the topic.

The unreasonable effectiveness of mathematics in science ...

Eugene Wigner

Well, if you know of a better 'ole, go to it.

Bruce Bairnsfather

What is now proved was once only imagined.

William Blake

As long as algebra and geometry proceeded along separate paths, their advance was slow and their applications limited.

But when these sciences joined company they drew from each other fresh vitality and thenceforward marched on at a rapid pace towards perfection.

Joseph Louis Lagrange.

Bussum, July 1985

Michiel Hazewinkel

PREFACE

The theory of submanifolds of a Kaehlerian manifold is one of the most interesting topics in differential geometry. According to the behaviour of the tangent bundle of a submanifold, with respect to the action of the almost complex structure of the ambient manifold, we have three typical classes of submanifolds: holomorphic submanifolds (see Ogiue [1]), totally real submanifolds (see Yano-Kon [1]) and CR- (Cauchy-Riemann) submanifolds. The notion of a CR-submanifold has been introduced by the author in [1] as follows:

Let N be an almost Hermitian manifold and let J be the almost complex structure of N. A real submanifold M of N is called a CR-submanifold if there exists a differentiable distribution D on M satisfying

(i) $J(D_x) = D_x$

and

(ii) $J(D_x^{\perp}) \subset T_xM^{\perp}$,

for each $x \in M$, where $D^{\perp}$ is the complementary orthogonal distribution to D and $T_xM^{\perp}$ is the normal space to M at x. Thus holomorphic submanifolds and totally real submanifolds are particular cases of CR-submanifolds. Moreover, each real hypersurface of N is a CR-submanifold which is neither a holomorphic submanifold nor a totally real submanifold.

The purpose of the book is to introduce the reader to the main problems of geometry of CR-submanifolds. In order to make it self-contained as much as possible we arrange in Chapter I most of the required background material.

Chapter II is devoted to the differential geometry of CR-submanifolds of almost Hermitian manifolds. The integrability of both of the distributions D and $D^{\perp}$ on a CR-submanifold are studied and there is obtained a class of linear connections with respect to which the CR-structure is parallel. Also, it is proven that CR-products do not exist in a sphere S^6.

In Chapter III we give results on some special classes

of CR-submanifolds of Kaehlerian manifolds: umbilical CR-submanifolds, normal CR-submanifolds, CR-products, Sasakian anti-holomorphic submanifolds. Also, the cohomology of CR-submanifolds is studied.

In Chapter IV we are concerned with CR-submanifolds of complex space forms. We first discuss the method of Riemannian fibre bundles in the geometry of CR-submanifolds. We include here various results on mixed foliate CR-submanifolds, CR-products, generic submanifolds and CR-submanifolds with semi-flat normal connection.

In Chapter V we show that the theory of CR-submanifolds of Kaehlerian manifolds initiated the study of new structures on submanifolds of several classes of manifolds. We only sketch a study of such structures in Sasakian manifolds and quaternion Kaehlerian manifolds.

The interrelation of the geometry of CR-submanifolds with the general theory of CR-manifolds is studied in Chapter VI. By means of some local f-structures with complemented frames we obtain results on pseudo-conformal mappings between CR-manifolds.

Finally, we show in Chapter VII an application of CR-structures to relativity. By Penrose correspondence we have an interesting passing from the geometry of a Minkowski space to the geometry of a CR-submanifold.

In concluding the preface I would like to express my sincere gratitude to Professor Michiel Hazewinkel for his valuable suggestions on both the content and the presentation of this book. Special thanks are due to Professor G. D. Ludden (Michigan State University) who edited the manuscript for language and terminology. I am indebted to Professor L. Verștraelen (Katholieke Universiteit Leuven) for his kind support during the printing of the book. My thanks go to all authors of books and articles, whose ideas we benefited in preparation the manuscript. I express my hearty thanks to my teachers at both Universities of Timişoara and Iaşi from whom I have learnt the differential geometry. Also, I should like to thank Dr. D. J. Larner for his patience and kind cooperation. It is a pleasant duty for me to acknowledge that D. Reidel Publishing Company took all possible care in the production of the book.

March 5, 1985 AUREL BEJANCU

Chapter I

DIFFERENTIAL-GEOMETRICAL STRUCTURES ON MANIFOLDS

§1. Linear Connections on a Manifold

Let N be a real n-dimensional connected differentiable manifold. Throughout the book all manifolds and morphisms are supposed to be differentiable of class C^∞. Denote by $\{\mathcal{U};\ x^h\}$ a system of coordinate neighborhoods on N, where $\mathcal{U}$ is a neighborhood and x^h are local coordinates in $\mathcal{U}$, with the indices h, i, j, k,... taking on values in the range $\{1, \ldots, n\}$. TN and $\mathcal{F}(N)$ are respectively the tangent bundle to N and the algebra of differentiable functions on N. Also we denote by $\Gamma(\mathcal{H})$ the module of differentiable sections of a vector bundle $\mathcal{H}$.

Then, for each $X \in \Gamma(TN)$ and $f \in \mathcal{F}(N)$ we define $Xf \in \mathcal{F}(N)$ by

$$Xf = X^h \frac{\partial f}{\partial x^h}, \tag{1.1}$$

where X^h are the local components of X with respect to the natural frame $\{\partial_h = \partial/\partial x^h\}$. In (1.1) and in the sequel, we make use of the Einstein convention, that is, repeated indices with one upper index and one lower index denote summation over their range.

A linear connection on N is a mapping

$$\nabla : \Gamma(TN) \times \Gamma(TN) \to \Gamma(TN); \quad (X, Y) \to \nabla_X Y,$$

satisfying the following conditions:

(i) $\qquad \nabla_{fX + Y}(Z) = f\nabla_X Z + \nabla_Y Z$

and

(ii) $\qquad \nabla_X(fY + Z) = f\nabla_X Y + (Xf)Y + \nabla_X Z,$

for any $f \in \mathcal{F}(N)$ and $X,Y,Z \in \Gamma(TN)$. The operator ∇_X is called the **covariant differentiation** with respect to X.

We define the covariant differentiation of a function

f with respect to X by

$$\nabla_X f = Xf.$$

Thus for any tensor field S of type (0, s) or (1, s) we define the covariant derivative $\nabla_X S$ of S with respect to X by

$$(\nabla_X S)(X_1,\dots,X_s) = \nabla_X(S(X_1,\dots,X_s)) - \\ - \sum_{i=1}^{s} \{S(X_1,\dots,\nabla_X X_i,\dots,X_s)\}, \qquad (1.2)$$

for any $X_i \in \Gamma(TN)$, $i = 1,\dots,s$. In a similar way we can define the covariant derivative of a tensor field of type (r, s), but for our purpose (1.2) is sufficient.

The tensor field S is said to be parallel with respect to the linear connection ∇ if we have

$$\nabla_X S = 0 \quad \text{for any} \quad X \in \Gamma(TN).$$

The torsion tensor T of a linear connection ∇ is a tensor field T of type (1, 2) defined by

$$T(X, Y) = \nabla_X Y - \nabla_Y X - [X, Y], \qquad (1.3)$$

for any $X, Y \in \Gamma(TN)$, where $[X, Y]$ is the Lie bracket of vector fields X and Y defined by

$$[X, Y](f) = X(Yf) - Y(Xf),$$

for any $f \in \mathcal{F}(N)$. A torsion-free connection is a linear connection with vanishing torsion tensor field. The curvature tensor R of a linear connection ∇ is a tensor field of type (1, 3) defined by

$$R(X, Y)Z = \nabla_X \nabla_Y Z - \nabla_Y \nabla_X Z - \nabla_{[X, Y]} Z,$$

for any $X, Y, Z \in \Gamma(TN)$.

§2. The Levi-Civita Connection

A tensor filed g of type (0, 2) is said to be a Riemannian metric on N if the following conditions are fulfilled:

(i) g is symmetric, i.e., $g(X, Y) = g(Y, X)$ for any $X, Y \in \Gamma(TN)$,

(ii) g is positive definite, i.e., $g(X, X) \geqslant 0$ for any $X \in \Gamma(TN)$ and $g(X, X) = 0$ if and only if $X = 0$. The manifold N endowed with a Riemannian metric g is called a Riemannian manifold.

The length of a vector X is denoted by $\|X\|$ and it is defined by $\|X\|^2 = g(X, X)$.

A linear connection ∇ on N is said to be a Riemannian connection if the Riemannian metric g is parallel with respect to ∇, i.e., by (1.2) we have

$$X(g(Y, Z)) = g(\nabla_X Y, Z) + g(Y, \nabla_X Z),$$

for all X, Y, Z $\in \Gamma(TN)$. The following Theorem is well known.

THEOREM 2.1. *On a Riemannian manifold there exists one and only one torsion-free Riemannian connection.*

The Riemannian connection whose existence and uniqueness are stated in this theorem is called the *Levi-Civita connection* and it is given by

$$2g(\nabla_X Y, Z) = X(g(Y, Z)) + Y(g(Z, X)) - Z(g(X, Y)) + $$
$$+ g([X, Y], Z) + g([Z, X], Y) - g([Y, Z], X), \quad (2.1)$$

for any X, Y, Z $\in \Gamma(TN)$.

From now on, in this section we let ∇ denote the Levi-Civita connection on N. Then by using the curvature tensor of ∇ defined by (1.4) we shall introduce some other tensor fields of interest for the geometry of a Riemannian manifold.

First, we define the *Riemannian curvature tensor* of type (0, 4) by

$$R(X, Y, U, V) = g(R(X, Y)U, V), \quad (2.2)$$

for any X, Y, U, V $\in \Gamma(TN)$. By using (1.4) and (2.2) it is easy to check the following formulas

$$\begin{aligned} &R(X, Y, U, V) + R(Y, X, U, V) = 0,\\ &R(X, Y, U, V) + R(X, Y, V, U) = 0,\\ &R(X, Y, U, V) = R(U, V, X, Y),\\ &R(X, Y, U, V) + R(Y, U, X, V) + R(U, X, Y, V) = 0. \end{aligned} \quad (2.3)$$

Next, we consider a local field of orthonormal frames $\{E_1, \ldots, E_n\}$ on N. Then

$$S(X, Y) = \sum_{i=1}^{n} \{g(R(E_i, X)Y, E_i)\}, \quad (2.4)$$

defines a global tensor field S of type (0, 2) called the *Ricci tensor field*. Using S we define a global scalar field ρ by

$$\rho = \sum_{i=1}^{n} \{S(E_i, E_i)\}, \tag{2.5}$$

called the <u>scalar curvature</u> of N.

For each plane γ spanned by orthonormal vectors X and Y in the tangent space T_xN, $x \in N$ we define the <u>sectional curvature</u> $K(\gamma)$ by

$$K(\gamma) = K_N(X \wedge Y) = g(R(X, Y)Y, X). \tag{2.6}$$

It is not difficult to see that $K(\gamma)$ is independent of the choice of the orthonormal basis $\{X, Y\}$ of γ. If $K(\gamma)$ is a constant for all planes γ in T_xN and for all points x of N, then N is called a <u>space of constant curvature</u> or a <u>real space form</u>. The following theorem due to Schur is well-known.

THEOREM 2.2. <u>Let N be a connected Riemannian manifold of dimension $n > 2$. If the sectional curvature $K(\gamma)$ depends only on the point x, then N is a real space form.</u>

We denote by N(c) a real space form of constant sectional curvature c. Then the curvature tensor of N(c) is given by

$$R(X, Y)Z = c(g(Y, Z)X - g(X, Z)Y), \tag{2.7}$$

for any $X, Y, Z \in \Gamma(TN)$.

The <u>Ricci curvature</u> with respect to a non-zero vector X is denoted by k(X) and it is defined by

$$k(X) = S(X, X)/g(X, X). \tag{2.8}$$

If the Ricci curvature at x is independent of the vector X, then the Ricci tensor is given by

$$S = kg, \tag{2.9}$$

where $k = k(X)$ for any $X \in T_xN$. If this is the case at every point of N and $n > 2$ then k is a constant on N. If $n = 2$, the Ricci tensor is always given by (2.9) and k is not necessarily a constant. A Riemannian manifold whose Ricci tensor is given by (2.9) is called an <u>Einstein space</u>.

If the curvature tensor R vanishes, that is, N is a space of zero curvature, we say that N is a <u>locally Euclidean space</u>.

Now, as it is well known, many applications of differential geometry are concerned with pseudo-Riemannian metrics (see Chapter VII for applications to the geometry of a Minkowski space). Let g be a tensor field of type (0, 2) on N. We say that g is of <u>constant index</u> if the dimension of the subspace $W \subset T_xN$ on which g is negative definite is the same for all $x \in N$. Then a non-degenerate and symmetric tensor field g of type (0, 2) on N is called a pseudo-Riemannian metric if it is of constant index. In this case we say that g has a signature of type (p, q) if the canonical form of g has p positive coefficients and q negative coefficients. The length of a vector X is defined as in the Riemannian case, that is, $\|X\|^2 = g(X, X)$.

Taking into account their importance in physics, pseudo-Riemannian manifolds have been intensively studied from different points of view. Several results in this field can be found in O'Neill [1].

§3. Submanifolds of a Riemannian Manifold

Let N be an n-dimensional Riemannian manifold and let M be an m-dimensional manifold immersed in N. Since we are dealing with a local study, we may assume that M is imbedded in N. Then M becomes a Riemannian submanifold of N with Riemannian metric induced by the Riemannian metric on N. Denote by $TM^{\perp}$ the normal bundle to M and by g both metrics on M and N. Also, we denote by $\tilde{\nabla}$ and ∇ the Levi-Civita connections on N and M respectively.

Then for any $X, Y \in \Gamma(TM)$ we have

$$\tilde{\nabla}_X Y = \nabla_X Y + h(X, Y), \tag{3.1}$$

where $h : \Gamma(TM) \times \Gamma(TM) \to \Gamma(TM^{\perp})$ is a normal bundle valued symmetric bilinear form on $\Gamma(TM)$. The equation (3.1) is called the Gauss formula and h is called the second fundamental form of M.

Now, for any $X \in \Gamma(TM)$ and $V \in \Gamma(TM^{\perp})$ we denote by $-A_VX$ and $\nabla^{\perp}_X V$ the tangential part and normal part of $\tilde{\nabla}_X V$ respectively. Then we have

$$\tilde{\nabla}_X V = -A_VX + \nabla^{\perp}_X V. \tag{3.2}$$

Thus for any $V \in \Gamma(TM^{\perp})$ we have a linear operator $A_V : \Gamma(TM) \to \Gamma(TM)$, satisfying

$$g(h(X, Y), V) = g(A_VX, Y). \tag{3.3}$$

The linear operator A_V is called the fundamental tensor of Weingarten with respect to the normal section V. The equation (3.2) is called the Weingarten formula.

The differential operator $\nabla^\perp$ defines a linear connection on the normal bundle $TM^\perp$ called the <u>normal connection</u> on M. By using this linear connection and the Levi-Civita connection on M we define the covariant derivative of h by

$$(\nabla_X h)(Y, Z) = \nabla^\perp_X(h(Y, Z)) - h(\nabla_X Y, Z) - \\ - h(Y, \nabla_X Z), \tag{3.4}$$

for all X, Y, Z $\in \Gamma(TM)$.

By a straightforward computation, using the Gauss and Weingarten formulas we obtain

$$\tilde{R}(X, Y)Z = R(X, Y)Z - A_{h(Y, Z)}X + A_{h(X, Z)}Y + \\ + (\nabla_X h)(Y, Z) - (\nabla_Y h)(X, Z), \tag{3.5}$$

for all X, Y, Z $\in \Gamma(TM)$, where R and $\tilde{R}$ are the curvature tensors of M and N respectively. Then from (3.5) and (3.3) follows the <u>Gauss equation</u>

$$g(\tilde{R}(X, Y)Z, U) = g(R(X, Y)Z, U) + \\ + g(h(X, Z), h(Y, U)) - \\ - g(h(Y, Z), h(X, U)), \tag{3.6}$$

for all X, Y, Z, U $\in \Gamma(TM)$. Also, taking the normal components of equation (3.5) we obtain the <u>Codazzi equation</u>

$$\{\tilde{R}(X, Y)Z\}^\perp = (\nabla_X h)(Y, Z) - (\nabla_Y h)(X, Z). \tag{3.7}$$

The curvature tensor $R^\perp$ of the normal connection $\nabla^\perp$ is defined by

$$R^\perp(X, Y)V = \nabla^\perp_X \nabla^\perp_Y V - \nabla^\perp_Y \nabla^\perp_X V - \nabla^\perp_{[X, Y]} V, \tag{3.8}$$

for all X, Y $\in \Gamma(TM)$ and V $\in \Gamma(TM^\perp)$. We define

$$[A_V, A_U] = A_V \circ A_U - A_U \circ A_V,$$

and by using the Gauss and Weingarten formulas obtain the <u>Ricci equation</u>

$$g(\tilde{R}(X, Y)V, U) = g(R^\perp(X, Y)V, U) + \\ + g([A_U, A_V]X, Y), \tag{3.9}$$

for any X, Y $\in \Gamma(TM)$ and V, U $\in \Gamma(TM^\perp)$.

If $R^{\perp} = 0$ we say that the normal connection on M is flat. A normal vector field V on M is called parallel if we have $\nabla^{\perp}_X V = 0$ for any $X \in \Gamma(TM)$.

THEOREM 3.1 (Chen [1]). Let M be an m-dimensional submanifold of an n-dimensional Riemannian manifold N. Then the normal connection $\nabla^{\perp}$ of M in N is flat if and only if there exist locally n - m mutually orthogonal unit normal vector fields V_i such that each of the V_i is parallel.

The submanifold M is said to be totally geodesic in N if its second fundamental form vanishes identically, i.e., $h = 0$ or equivalently $A_V = 0$, for any $V \in \Gamma(TM^{\perp})$. If, for a normal section V on M, we have $A_V = \alpha I$, where α is a differentiable function and I is the identity morphism on $\Gamma(TM)$, then M is called totally umbilical with respect to the normal section V. If M is totally umbilical with respect to any normal section we say that M is totally umbilical.

Let $\{E_1, \ldots, E_m\}$ be an orthonormal basis in T_xM. Then

$$\mathrm{Tr}(h) = \sum_{i=1}^{m} \{h(E_i, E_i)\},$$

is independent of the basis and it is called the trace of h. By means of Tr(h) we define the mean curvature vector H of M by

$$H = \frac{1}{m}\,\mathrm{Tr}(h). \tag{3.10}$$

If $H = 0$ on M we say that M is a minimal submanifold of N. Moreover, we obtain that M is totally umbilical if and only if we have

$$h(X, Y) = g(X, Y)H, \tag{3.11}$$

for any $X, Y \in \Gamma(TM)$.

Finally, M is called a submanifold with parallel second fundamental form if we have $\overline{\nabla} h = 0$.

§4. Distributions on a Manifold.

An m-dimensional distribution on a manifold N is a mapping D defined on N, which assignes to each point x of N an m-dimensional linear subspace D_x of T_xN. A vector field X on N belongs to D if we have $X_x \in D_x$ for each $x \in N$. When this happens we write $X \in \Gamma(D)$. The distribution D is said to be differentiable if for any $x \in N$ there exist m differentiable linearly independent vector fields $X_i \in \Gamma(D)$ in a neighborhood of x. From now on, all distributions are supposed to be differentiable of class C^∞.

The distribution D is said to be involutive if for all vector fields $X, Y \in \Gamma(D)$ we have $[X, Y] \in \Gamma(D)$. A submanifold M of N is said to be an integral manifold of D if for every point $x \in M$, D_x coincides with the tangent space to M at x. If there exists no integral manifold of D which contains M, then M is called a maximal integral manifold or a leaf of D. The distribution D is said to be integrable if for every $x \in N$ there exists an integral manifold of D containing x.

Then the well known theorem of Frobenius is stated as follows.

THEOREM 4.1. Each involutive distribution is integrable. Moreover, through every point $x \in N$ there passes a unique maximal integral manifold of D and every other integral manifold containing x is an open submanifold of the maximal one.

Let ∇ be a linear connection on N. The distribution D is said to be parallel with respect to ∇ if we have

$$\nabla_X Y \in \Gamma(D) \quad \text{for any} \quad X \in \Gamma(TN) \quad \text{and} \quad Y \in \Gamma(D). \tag{4.1}$$

Now, suppose N is endowed with two complementary distributions D and $\tilde{D}$, i.e., we have $TN = D \oplus \tilde{D}$. Denote by P and Q the projection morphisms of TN to D and $\tilde{D}$ respectively.

THEOREM 4.2. All the linear connections with respect to which both distributions D and $\tilde{D}$ are parallel, are given by

$$\nabla_X Y = P\mathring{\nabla}_X PY + Q\mathring{\nabla}_X QY + PS(X, PY) + QS(X, QY), \tag{4.2}$$

for any $X, Y \in \Gamma(TN)$, where $\mathring{\nabla}$ and S are, respectively, an arbitrary linear connection on N and an arbitrary tensor field of type (1, 2) on N.

Proof. Suppose $\mathring{\nabla}$ is an arbitrary linear connection on N. Then any linear connection ∇ on N is given by

$$\nabla_X Y = \mathring{\nabla}_X Y + S(X, Y), \tag{4.3}$$

for any $X, Y \in \Gamma(TN)$, where S is an arbitrary tensor field of type (1, 2) on N. The distributions D and $\tilde{D}$ are both parallel with respect to ∇ if and only if we have

$$Q(\nabla_X Y) = 0 \quad \text{and} \quad P(\nabla_X Z) = 0, \tag{4.4}$$

for any $X \in \Gamma(TN)$, $Y \in \Gamma(D)$ and $Z \in \Gamma(\tilde{D})$. From (4.3) and (4.4) it follows that D and $\tilde{D}$ are parallel with respect to ∇ if and only if we have

$$Q\overset{\circ}{\nabla}_X Y + QS(X, Y) = 0 \quad \text{and} \quad P\overset{\circ}{\nabla}_X Z + PS(X, Z) = 0. \tag{4.5}$$

Thus, taking account of (4.5) in (4.3), we obtain (4.2).

Now, by means of the projection morphisms P and Q we define a tensor field F of type (1, 1) on N by

$$FX = PX - QX, \tag{4.6}$$

for any $X \in \Gamma(TN)$. By direct computation follows $F^2 = I$. We say that F defines an almost product structure on N. The covariant derivative of F is defined by

$$(\nabla_X F)Y = \nabla_X FY - F(\nabla_X Y), \tag{4.7}$$

for all $X, Y \in \Gamma(TN)$. We say that the almost product structure F is parallel with respect to the linear connection ∇ if we have $\nabla_X F = 0$ for each $X \in \Gamma(TN)$. Then by using (4.4), (4.6) and (4.7) we obtain

THEOREM 4.3. *Both distributions D and $\tilde{D}$ are parallel with respect to ∇ if and only if the almost product structure F is parallel with respect to ∇.*

Now suppose N is a Riemannian manifold endowed with two complementary orthogonal distributions D and $D^\perp$. Denote by ∇ the Levi-Civita connection on N. Then we have

THEOREM 4.4. *Both distributions D and $D^\perp$ are parallel with respect to Levi-Civita connection ∇ if and only if they are integrable and their leaves are totally geodesic in N.*

Proof. Suppose both distributions D and $D^\perp$ are parallel with respect to ∇. Then since ∇ is a torsion free linear connection, we have

$$[X, Y] = \nabla_X Y - \nabla_Y X \in \Gamma(D) \quad \text{for any } X, Y \in \Gamma(D)$$

and

$$[U, V] = \nabla_U V - \nabla_V U \in \Gamma(D^\perp) \quad \text{for any } U, V \in \Gamma(D^\perp).$$

Thus, by the Frobenius Theorem, both distributions D and $D^\perp$ are integrable. Now, let M be a leaf of D and denote by h the second fundamental form of the immersion of M in N. Then by the Gauss formula we have

$$h(X, Y) = \nabla_X Y - \nabla'_X Y, \tag{4.8}$$

for any $X, Y \in \Gamma(TM)$, where ∇' is the Levi-Civita connection on M. Since $\nabla_X Y \in \Gamma(TM)$ and $h(X, Y) \in \Gamma(D^\perp)$, by (4.8) we obtain $h = 0$, that is M is totally geodesic in N. In a similar way it follows that each leaf of $D^\perp$ is totally geodesic in N.

Conversely, suppose D and $D^\perp$ be integrable and their leaves are totally geodesic in N. Then, by using (3.1) for the immersions of the leaves of D and $D^\perp$ in N, we obtain

$$\nabla_X Y \in \Gamma(D) \quad \text{for any} \quad X, Y \in \Gamma(D)$$

and

$$\nabla_U V \in \Gamma(D^\perp) \quad \text{for any} \quad U, V \in \Gamma(D^\perp).$$

Since g is parallel with respect to ∇ we have

$$g(\nabla_U Y, V) = -g(Y, \nabla_U V) = 0$$

and

$$g(\nabla_X V, Y) = -g(V, \nabla_X Y) = 0,$$

for any $X, Y \in \Gamma(D)$ and $U, V \in \Gamma(D^\perp)$. Thus both distributions D and $D^\perp$ are parallel on N. The proof is complete.

From Theorem 4.4 it follows that if N is endowed with two complementary orthogonal distributions D and $D^\perp$ that are parallel with respect to the Levi-Civita connection, then N is locally a Riemannian product $M \times M^\perp$, where M and $M^\perp$ are leaves of D and respectively $D^\perp$.

Finally, taking account of the fact that g is parallel with respect to ∇, we obtain

PROPOSITION 4.1. The distribution D is parallel with respect to the Levi-Civita connection ∇ if and only if the complementary orthogonal distribution $D^\perp$ is parallel with respect to ∇.

§5. Kaehlerian Manifolds

Let N be a real differentiable manifold. An almost complex

structure on N is a tensor field J of type (1, 1) on N such that at every point $x \in N$ we have $J^2 = -I$, where I denotes the indentity transformation of T_xN. A manifold N endowed with an almost complex structure is called an almost complex manifold. Every almost complex manifold is of even dimension and is orientable.

Now suppose (N, J) is an almost complex manifold. Then we define the torsion tensor of J or the Nijenhuis tensor of J by

$$[J, J](X, Y) = [JX, JY] - [X, Y] - J[JX, Y] - J[X, JY] \tag{5.1}$$

for any $X, Y \in \Gamma(TN)$. If the torsion tensor of J vanishes identically on N then we say that J is a complex structure on N and N becomes a complex manifold. From (5.1) we have

$$[J, J](JX, Y) = -J([J, J](X, Y)) = [J, J](X, XY). \tag{5.2}$$

A Hermitian metric on an almost complex manifold N is a Riemannian metric g satisfying

$$g(JX, JY) = g(X, Y),$$

for any $X, Y \in \Gamma(TN)$. An almost complex manifold (resp. a complex manifold) endowed with a Hermitian metric is called an almost Hermitian manifold (resp. a Hermitian manifold). Every almost complex manifold with a Riemannian metric g' admits a Hermitian metric. In fact, if we take

$$g(X, Y) = g'(X, Y) + g'(JX, JY),$$

it is easily seen that g is a Hermitian metric on N.

The 2-form Ω of a Hermitian manifold N is defined by

$$\Omega(X, Y) = g(X, JY), \quad \text{for any } X, Y \in \Gamma(TN). \tag{5.3}$$

By a direct computation it follows

$$3d\Omega(X, Y, Z) = g((\tilde{\nabla}_X J)Y, Z) + g(\tilde{\nabla}_Y J)Z, X) + g((\tilde{\nabla}_Z J)X, Y), \tag{5.4}$$

for all $X, Y, Z \in \Gamma(TN)$, where $\tilde{\nabla}$ is the Levi-Civita connection on N.

We say that N is a **Kaehlerian manifold** if its 2-form Ω is closed, i.e., we have $d\Omega = 0$. It is not difficult to prove that a Hermitian manifold is a Kaehlerian manifold if and only if the almost complex structure J is parallel with respect to $\tilde{\nabla}$, i.e., we have $\tilde{\nabla}_X J = 0$ for any $X \in \Gamma(TN)$. An almost Hermitian manifold is called a <u>nearly Kaehlerian manifold</u> if we have

$$(\tilde{\nabla}_X J)X = 0, \quad \text{for any} \quad X \in \Gamma(TN). \tag{5.5}$$

From (5.5) we obtain that N is a nearly Kaelerian manifold if and only if we have

$$(\tilde{\nabla}_X J)Y + (\tilde{\nabla}_Y J)X = 0, \quad \text{for any} \quad X, Y \in \Gamma(TN). \tag{5.6}$$

PROPOSITION 5.1. *Let N be a nearly Kaelerian manifold. Then the Nijenhuis tensor of J is given by*

$$[J, J](X, Y) = 4J(\tilde{\nabla}_Y J)X, \tag{5.7}$$

for any $X, Y \in \Gamma(TN)$.

Proof. Taking account of the fact that $\tilde{\nabla}$ is a torsion free connection on N, from (5.1) we obtain

$$[J, J](X, Y) = (\tilde{\nabla}_{JX}J)Y - (\tilde{\nabla}_{JY}J)X + J((\tilde{\nabla}_Y J)X) - $$
$$- J((\tilde{\nabla}_X J)Y). \tag{5.8}$$

By using (5.6) we have

$$(\tilde{\nabla}_{JY}J)X = -(\tilde{\nabla}_X J)JY = \tilde{\nabla}_X Y + J(\tilde{\nabla}_X JY) = J((\tilde{\nabla}_X J)Y)$$

Thus (5.8) becomes

$$[J, J](X, Y) = 2(\tilde{\nabla}_Y X + J(\tilde{\nabla}_Y JX) - \tilde{\nabla}_X Y - J(\tilde{\nabla}_X JY))$$
$$= 2\{J(\tilde{\nabla}_Y JX - J(\tilde{\nabla}_Y X)) - J(\tilde{\nabla}_X JY - J(\tilde{\nabla}_X Y))\}$$
$$= 2J((\tilde{\nabla}_Y J)X - (\tilde{\nabla}_X J)Y) = 4J((\tilde{\nabla}_Y J)X),$$

which proves our assertion.

The curvature tensor $\tilde{R}$ of a Kaehlerian manifold N satisfies

$$\tilde{R}(X, Y)J = J\tilde{R}(X, Y) \quad \text{and} \quad \tilde{R}(JX, JY) = \tilde{R}(X, Y), \tag{5.9}$$

for any $X, Y \in \Gamma(TN)$.

Now we consider a plane γ invariant by the almost complex structure J. In this case we can choose a basis $\{X, JX\}$ in γ, where X is a unit vector in γ. Then the sectional curvature $K(\gamma)$ is denoted by $H(X)$ and it is called the <u>holomorphic sectional curvature</u> of N determined by the unit vector X. Thus, by using (2.6), we have

$$H(X) = g(\tilde{R}(X, JX)JX, X) \tag{5.10}$$

If $H(X)$ is a constant for all unit vectors in T_xN and for all points $x \in N$, then N is called a space of constant holomorphic sectional curvature.

THEOREM 5.1. Let N be a connected Kaehlerian manifold of complex dimension $n \geqslant 2$. If the holomorphic sectional curvature $H(X)$ depends only on $x \in N$, then N is a space of constant holomorphic sectional curvature.

A Kaehlerian manifold of constant holomorphic sectional curvature is called a complex space form.

THEOREM 5.2. Any two simply connected complete complex space forms of constant holomorphic sectional curvature c are holomorphically isometric to each other.

For the proofs of Theorem 5.1 and 5.2 see Kobayashi and Nomizu II, p. 168 and 170.

The curvature tensor of a complex space form N of constant holomorphic sectional curvature c is given by

$$\tilde{R}(X, Y)Z = \frac{c}{4}\{g(Y, Z)X - g(X, Z)Y + g(Z, JY)JX - g(Z, JX)JY + 2g(X, JY)JZ\}, \qquad (5.11)$$

for any $X, Y, Z \in \Gamma(TN)$.

The holomorphic bisectional curvature for the pair of unit vectors $\{X, Y\}$ is given by

$$H_B(X \wedge Y) = g(\tilde{R}(X, JX)JY, Y). \qquad (5.12)$$

A Typical Example of a Nearly Kaehlerian Manifold

Let C be the Cayley division algebra generated by $\{e_0 = 1, e_i, 1 \leqslant i \leqslant 7\}$, over R, and C_+ the subspace of C consisting of all purely imaginary Cayley numbers. We may identify C_+ with a 7-dimensional Euclidean space R^7 with the canonical inner product (,). The automorphism group of C_+ is, by definition, the compact simple Lie group G_2. Moreover, the inner product (,) is invariant under the action of G_2 and hence G_2 may be considered as a subgroup of SO(7). A vector cross product for vectors in $R^7 = C_+$ is defined by

$$x \times y = (x, y)e_0 + xy, \quad x, y \in C_+.$$

Then the multiplication table is given by

$e_j \times e_k =$

j/k	1	2	3	4	5	6	7
1	0	e_3	$-e_2$	e_5	$-e_4$	e_7	$-e_6$
2	$-e_3$	0	e_1	e_6	$-e_7$	$-e_4$	e_5
3	e_2	$-e_1$	0	$-e_7$	$-e_6$	e_5	e_4
4	$-e_5$	$-e_6$	e_7	0	e_1	e_2	$-e_3$
5	e_4	e_7	e_6	$-e_1$	0	$-e_3$	$-e_2$
6	$-e_7$	e_4	$-e_5$	$-e_2$	e_3	0	e_1
7	e_6	$-e_5$	$-e_4$	e_3	e_2	$-e_1$	0

Considering the unit sphere S^6 as $\{x \in C_+;\ (x, x) = 1\}$, an almost complex structure J on S^6 is defined by

$$J_x U = x \times U,$$

where $x \in S^6$ and $U \in T_x(S^6)$. The almost complex structure J together with the induced metric g on S^6 from the inner product (,) on $R^7 = C_+$ gives rise to a nearly Kaehlerian structure on S^6. The group G_2 acts on S^6 transitively as the group of automorphisms of the nearly Kaehlerian structure (J, g), (see Fukami-Ishihara [1]).

Examples of Kaehlerian Manifolds

(1). The complex n-space C^n with the metric

$$ds^2 = \sum_{a=1}^{n} dz^a\, d\bar{z}^a,$$

where $(z^1,\ldots,z^n)$ is the natural coordinate system, is a complete, flat Kaehlerian manifold with fundamental 2-form

$$\Omega = -i \sum_{a=1}^{n} dz^a \wedge d\bar{z}^a.$$

(2). Let CP^n be the n-dimensional complex projective space. Then for any positive number c, CP^n carries a Kaehlerian metric of constant holomorphic sectional curvature c given by

$$ds^2 = \frac{4}{c}\,\frac{(1+\Sigma z^a\bar{z}^a)(\Sigma dz^a d\bar{z}^a) - (\Sigma \bar{z}^a dz^a)(\Sigma z^a d\bar{z}^a)}{(1+\Sigma z^a\bar{z}^a)^2},$$

where $(z^1,\ldots,z^n)$ is an inhomogeneous coordinate system of CP^n.

(3). Let D^n be the unit open ball in C^n, i.e.,

$$D^n = \{(z^1,\ldots,z^n);\quad \Sigma z^a\bar{z}^a < 1\}.$$

Then for any negative number c, D^n carries a Kaelerian metric of constant holomorphic sectional curvature c given by

$$ds^2 = -\frac{4}{c}\,\frac{(1-\Sigma z^a\bar{z}^a)(\Sigma dz^a d\bar{z}^z) - (\Sigma \bar{z}^a dz^a)(\Sigma z^a d\bar{z}^a)}{(1-\Sigma z^a\bar{z}^a)^2}.$$

For other examples and details of these examples see Kobayashi and Nomizu II, p.159.

By using Theorem 5.2 and these examples we obtain that a simply connected complete Kaehlerian manifold of constant holomorphic sectional curvature c can be identified with the complex projective space CP^n, the open unit ball D^n in C^n or with C^n, according as $c > 0$, $c < 0$ or $c = 0$ respectively.

§6. Sasakian Manifolds

Let N be a real (2n + 1)-dimensional differentiable manifold and ϕ, ξ, and η be a tensor field of type (1, 1), a vector field and a 1-form respectively on N satisfying

$$\phi^2 = -I + \eta\otimes\xi;\ \phi\xi = 0;\ \eta(\phi X) = 0;\ \eta(\xi) = 1, \qquad (6.1)$$

for any $X \in \Gamma(TN)$. Then N is called an almost contact manifold and (ϕ, ξ, η) the almost contact structure on N.

Now, suppose on N is given a Riemannian metric tensor field g which satisfies the equations

$$g(\phi X, \phi Y) = g(X, Y) - \eta(X)\cdot\eta(Y) \qquad (6.2)$$

and

$$\eta(X) = g(X, \xi), \qquad (6.3)$$

for any $X, Y \in \Gamma(TN)$. Then N is called an almost contact metric manifold and (ϕ, ξ, η, g) the almost contact metric structure. If we have

$$d\eta(X, Y) = g(X, \phi Y), \tag{6.4}$$

for all $X, Y \in \Gamma(TN)$ then we say that N is a contact metric manifold.

The Nijenhuis tensor field of ϕ is defined by

$$[\phi, \phi](X, Y) = [\phi X, \phi Y] + \phi^2[X, Y] - \phi[X, \phi Y] - \phi[\phi X, Y], \tag{6.5}$$

for all $X, Y \in \Gamma(TN)$. If N is an almost contact metric manifold and the Nijenhuis tensor of ϕ satisfies

$$[\phi, \phi] + 2d\eta \otimes \xi = 0, \tag{6.6}$$

then we say that N is a normal almost contact manifold.

A normal contact manifold is called a Sasakian manifold. It is known (see Blair [3]) that a contact metric structure is Sasakian if and only if we have

$$(\tilde{\nabla}_X \phi) Y = g(X, Y)\xi - \eta(Y)X, \tag{6.7}$$

for any $X, Y \in \Gamma(TN)$, where $\tilde{\nabla}$ is the Levi-Civita connection on N with respect to the Riemannian metric g. Also, on a Sasakian manifold we have

$$\tilde{\nabla}_X \xi = -\phi X, \quad \text{for any } X \in \Gamma(TN). \tag{6.8}$$

We denote by $\{\xi\}^\perp$ the complementary orthogonal distribution to the 1-dimensional distribution spanned by ξ on N. A plane section in the tangent space T_xN is said to be a ϕ-section if it is spanned by X and ϕX where $X \in \{\xi\}^\perp_x$. The sectional curvature $K(\gamma)$ determined by a ϕ-section γ is called a ϕ-sectional curvature. If a Sasakian manifold has a ϕ-sectional curvature c which does not depend on the ϕ-section at each point, then c is a constant on the manifold. A Sasakian space form $N(c)$ is a Sasakian manifold of constant ϕ-sectional curvature c.

The curvature tensor of a Sasakian space form $N(c)$ is given by

$$\begin{aligned}\tilde{R}(X, Y)Z = {} & \frac{c+3}{4}\{g(Y, Z)X - g(X, Z)Y\} - \\ & - \frac{c-1}{4}\{\eta(Y)\cdot\eta(Z)X - \eta(X)\cdot\eta(Z)Y + g(Y, Z)\cdot\eta(X)\zeta - \\ & - g(X, Z)\cdot\eta(Y)\xi - g(\phi Y, Z)\phi X + g(\phi X, Z)\phi Y + \\ & + 2g(\phi X, Y)\phi Z\}, \quad \text{for any } X, Y, Z \in \Gamma(TN).\end{aligned} \tag{6.9}$$

Examples of Sasakian Manifolds

(1) Let S^{2n+1} be a (2n+1)-dimensional unit sphere, i.e.,

$$S^{2n+1} = \{z \in C^{n+1};\ |z| = 1\}.$$

Denote by J the almost complex structure on C^{n+1} and define $\xi_z = Jz$ for any $z \in S^{2n+1}$. Next we consider the orthogonal projection

$$\pi : T_z(C^{n+1}) \to T_z(S^{2n+1}),$$

and define $\phi = \pi \circ J$. Thus we have a Sasakian structure (ϕ, ξ, η, g) on S^{2n+1}, where η is a 1-form dual to ξ and g is the standard metric tensor field on S^{2n+1}. Moreover, S^{2n+1} is of constant ϕ-sectional curvature 1, that is a Sasakian space form N(1).

(2) Let E^{2n+1} be an Euclidean space with cartesian coordinates $(x^1,\dots,x^n, y^1,\dots,y^n, z)$. Then we define a Sasakian structure (ϕ, ξ, η, g) on E^{2n+1} by the following formulas

$$\xi = (0,\dots,0, 2), \quad 2\eta = (-y^1,\dots,-y^n, 0,\dots,0, 1),$$

$$g_{AB} = \begin{bmatrix} \frac{1}{4}(\delta_{ij} + y^i y^j) & 0 & -\frac{1}{4} y^i \\ 0 & \frac{1}{4}\delta_{ij} & 0 \\ -\frac{1}{4} y^i & 0 & \frac{1}{4} \end{bmatrix}$$

and

$$\phi^A_B = \begin{bmatrix} 0 & \delta^i_j & 0 \\ -\delta^i_j & 0 & 0 \\ 0 & y^j & 0 \end{bmatrix}$$

Thus E^{2n+1} is a Sasakian space form of constant ϕ-sectional curvature c = -3 and it is denoted by $E^{2n+1}(-3)$.

More examples and results on the geometry of Sasakian manifolds can be found in Blair [3].

7. Quaternion Kaehlerian Manifolds

Let N be a real n-dimensional differentiable manifold. Suppose there exists a vector bundle V consisting of tensors of type (1, 1) over N satisfying the following condition. In any coordinate neighborhood $\mathcal{U}$ of N, there is a local basis $\{J_1, J_2, J_3\}$ of V such that we have

$$\begin{cases} (J_1)^2 = (J_2)^2 = (J_3)^2 = -I, \\ J_1 \circ J_2 = -J_2 \circ J_1 = J_3, \end{cases} \tag{7.1}$$

where I is the identity tensor field of type (1, 1) in N. Such a local basis $\{J_1, J_2, J_3\}$ is called a canonical local basis of the vector fundle V in $\mathcal{U}$. The manifold N is called an almost quaternion manifold. From the definition we obtain that the almost quaternion manifold is of real dimension $n = 4m$ $(m \geqslant 1)$ and each fibre of the vector bundle is 3-dimensional.

Let N be an almost quaternion manifold and $\mathcal{U}$, $\tilde{\mathcal{U}}$ be two coordinate neighborhoods such that $\mathcal{U} \cap \tilde{\mathcal{U}} \neq \emptyset$. We consider the canonical local basis $\{J_1, J_2, J_3\}$ and $\{\tilde{J}_1, \tilde{J}_2, \tilde{J}_3\}$ in $\mathcal{U}$ and $\tilde{\mathcal{U}}$ respectively. Then we have

$$\tilde{J}_a = \sum_{b=1}^{3} S_a^b J_b \quad (a = 1, 2, 3), \tag{7.2}$$

where S_a^b are differentiable functions on $\mathcal{U} \cap \tilde{\mathcal{U}}$. By using (7.1) for both basis we obtain that at each point the coefficients $S_b^a(x)$ define an element of the proper orthogonal group SO(3) of dimension 3. By means of this result it follows that an almost quaternion manifold is orientable.

Now we suppose that N is an almost quaternion manifold endowed with a Riemannian metric g satisfying

$$g(X, \phi Y) + g(\phi X, Y) = 0, \tag{7.3}$$

for any $X, Y \in \Gamma(TN)$ and any local section ϕ of the vector bundle V. In this case we say that N is an almost quaternion metric manifold. Thus if $\{J_1, J_2, J_3\}$ is a canonical basis of V, each of J_1, J_2, J_3 is almost Hermitian with respect to g. Define three local 2-forms Ω_a on $\mathcal{U}$ by

$$\Omega_a(X, Y) = g(X, J_aY), \quad a = 1, 2, 3, \tag{7.4}$$

for all vector fields X, Y on $\mathcal{U}$. Then Ω defined by

$$\Omega = \sum_{a=1}^{3} \Omega_a \wedge \Omega_a \tag{7.5}$$

is a 4-form globally defined on N.

Next, we denote by $\tilde{\nabla}$ the Levi-Civita connection on the almost quaternion metric manifold N. We say that N is a quaternion Kaehlerian manifold if, for each local section ϕ of V and vector field X on N, $\tilde{\nabla}_X\phi$ is also a section of V. It is not difficult to see that N is a quaternion Kaehlerian manifold if and only if we have

$$\tilde{\nabla}_X J_a = \sum_{b=1}^{3} Q_{ab}(X) J_b, \quad a = 1, 2, 3, \tag{7.6}$$

for any $X \in \Gamma(TN)$, where Q_{ab} are certain local 1-forms on N such that $Q_{ab} + Q_{ba} = 0$. By using (7.4), (7.5), and (7.6) we obtain that an almost quaternion metric manifold is a quaternion Kaehlerian manifold if and only if $\tilde{\nabla}\Omega = 0$.

Now, let N be a quaternion Kaehlerian manifold and X be a unit vector tangent to N at x. We denote by $\gamma(X)$ the 4-dimensional vector subspace of T_xN spanned by $\{X, J_1X, J_2X, J_3X\}$ and call it the quaternion 4-space determined by X. A quaternion plane is a 2-dimensional vector subspace of a quaternion 4-space. The sectional curvature for a quaternion plane is called a quaternion sectional curvature. If the quaternion sectional curvature is a constant c for all quaternion planes and for all points x of N we say that N is a quaternion space form and denote it by N(c). The curvature tensor $\tilde{R}$ of a quaternion space form N(c) is given by

$$\tilde{R}(X, Y)Z = \frac{c}{4}\{g(Y, Z)X - g(X, Z)Y + \sum_{a=1}^{3}\{g(J_aY, Z)J_aX - g(J_aX, Z)J_aY + 2g(X, J_aY)J_aZ\}\}, \tag{7.7}$$

for any $X, Y, Z \in \Gamma(TN)$.

Chapter II

CR-SUBMANIFOLDS OF ALMOST HERMITIAN MANIFOLDS

§1. CR-submanifolds and CR-structures

Let N be a n-dimensional almost Hermitian manifold with almost complex structure J and with Hermitian metric g. Let M be a real m-dimensional Riemannian manifold isometrically immersed in N.

The differential geometry of M depends on the behaviour of the tangent bundle of M relative to the action of the almost complex structure J. Thus M is called a complex (holomorphic) submanifold if T_xM is invariant by J, i.e., we have

$$J(T_xM) = T_xM, \quad \text{for each} \quad x \in M.$$

Also, we say that M is a totally real (anti-invariant) submanifold of N if we have

$$J(T_xM) \subset T_xM^{\perp}, \quad \text{for each} \quad x \in M.$$

These two classes of submanifolds have been extensively investigated in the last decade from different viewpoints. The fundamental results on the geometry of totally real submanifolds can be found in Yano-Kon [1]. Also, a survey of the principal results on the geometry of complex submanifolds is given by Ogiue in [1].

In 1978 we initiated in [1] a study of the differential geometry of a new class of submanifolds situated between the above two classes, called CR-submanifolds. More precisely, M is said to be a CR-submanifold of N if there exists a differentiable distribution

$$D : x \to D_x \subset T_xM,$$

on M satisfying the following conditions:

(i) D is holomorphic, i.e., $J(D_x) = D_x$ for each $x \in M$,

(ii) the complementary orthogonal distribution

$$D^{\perp} : x \to D_x^{\perp} \subset T_xM,$$

is anti-invariant, i.e., $J(D_x^{\perp}) \subset T_xM^{\perp}$, for each $x \in M$.

We denote by p the complex dimension of the distribution

D and by q the real dimension of the distribution $D^{\perp}$. Then for q = 0 (resp. p = 0) a CR-submanifold becomes a complex submanifold (resp. totally real submanifold). If $q = \dim T_xM^{\perp}$ the CR-submanifold is called an anti-holomorphic submanifold. A proper CR-submanifold is a CR-submanifold which is neither a complex submanifold nor a totally real submanifold.

Each real hypersurface M of N ($n \geq 2$) is a proper CR-submanifold. In fact we define

$$D^{\perp} : x \to D_x^{\perp} = J(T_xM^{\perp})$$

and take D as the complementary orthogonal distribution to $D^{\perp}$ in TM. Thus M is endowed with a pair of distributions $(D, D^{\perp})$ satisfying the conditions of the definition of a CR-submanifold. Moreover, we have $\dim_R D_x^{\perp} = 1$ and $\dim_C D_x = n - 1$. Hence, M is a proper CR-submanifold.

Remark 1.1. It is easily seen that on a CR-submanifold the distribution D (resp. $D^{\perp}$) is the maximal distribution invariant by J (resp. anti-invariant by J), i.e., if D' (resp. D") is an invariant (resp. anti-invariant) distribution on M, then we have $D'_x \subset D_x$ (resp. $D''_x \subset D_x^{\perp}$), for each $x \in M$.

Now let M be an arbitrary Riemannian manifold isometrically immersed in an almost Hermitian manifold N. For each vector field X tangent to M we put

$$JX = \phi X + \omega X, \tag{1.1}$$

where ϕX and ωX are respectively the tangent part and the normal part of JX. Also, for each vector field V normal to M we put

$$JV = BV + CV, \tag{1.2}$$

where BV and CV are respectively the tangent part and the normal part of JV.

THEOREM 1.1. The submanifold M of N is a CR-submanifold if and only if we have

$$\operatorname{rank}(\phi) = \text{constant}, \tag{1.3}$$

and

$$\omega \circ \phi = 0. \tag{1.4}$$

Proof. Suppose M is a CR-submanifold of an almost Hermitian manifold N. Denote by P and Q respectively the projection morphisms of TM to D and $D^{\perp}$. Then we have

$$\phi X = JPX \tag{1.5}$$

and

$$\omega X = JQX, \quad \text{for any} \quad X \in \Gamma(TM). \tag{1.6}$$

Thus from (1.5) it follows rank(ϕ) = 2p and from (1.6), taking account of (1.5), (1.4) follows.

Conversely, suppose (1.3) and (1.4) are satisfied. We define the distribution D by

$$D_x = \text{Im. } \phi_x, \quad \text{for each} \quad x \in M.$$

Clearly, D is an invariant distribution, since for each $X = \phi Y \in \Gamma(D)$ we have

$$JX = J\phi Y = \phi^2 Y + (\omega\circ\phi)Y = \phi^2 Y \in \Gamma(\text{Im. } \phi) = \Gamma(D).$$

Denote by $D^\perp$ the complementary orthogonal distribution to D in TM. Then $D^\perp$ is an anti-invariant distribution. In fact, for any $X \in \Gamma(D^\perp)$ and $Y = U + W$ where $U \in \Gamma(D)$ and $W \in \Gamma(D^\perp)$ we have

$$g(JX, Y) = -g(X, JU + JW) = -g(X, JW) =$$
$$= -g(X, \phi W) = 0,$$

since $\phi W \in \Gamma(D)$. Thus M is a CR-submanifold of N and the proof is complete.

THEOREM 1.2. _The submanifold M of N is a CR-submanifold if and only if we have_

$$\text{rank}(B) = \text{constant}, \tag{1.7}$$

and

$$\phi\circ B = 0. \tag{1.8}$$

Proof. Suppose M is a CR-submanifold. First we see that $\text{Im.}B_x \subset D_x^\perp$ for each $x \in M$. In fact, we have

$$g(BV, Y) = g(JV, Y) = -g(V, JY) = 0.$$

for any $Y \in \Gamma(D)$ and $V \in \Gamma(TM^\perp)$. On the other hand, we have $D_x^\perp \subset \text{Im.}B_x$. Indeed, if we take $U \in D_x^\perp$ then $JU \in T_xM^\perp$ and we obtain

$$-U = J^2 U = BJU + CJU.$$

Hence $U = -BJU \in \text{Im.}B_x$. Thus we have $D^\perp = \text{Im.}B$, which implies rank(B) = constant. Next, for each $V \in \Gamma(TM^\perp)$ we have

$$(J \circ B)V = (\phi \circ B)V + (\omega \circ B)V.$$

Hence $(\phi \circ B)V = 0$ since both $(J \circ B)V$ and $(\omega \circ B)V$ are normal to M. Thus (1.8) is proven.

Conversely, suppose (1.7) and (1.8) be satisfied. Then we define the distribution $D^{\perp}$ by $D_x^{\perp} = \mathrm{Im}.B_x$. First we note that $D^{\perp}$ is an anti-invariant distribution. In fact, for each $X \in \Gamma(D^{\perp})$ and $Y \in \Gamma(TM)$ we have

$$g(JX, Y) = g((J \circ BV, Y) = g((\phi \circ B)V, Y) = 0,$$

by (1.8).
Next, the complementary orthogonal distribution D to $D^{\perp}$ in TM is a holomorphic distribution. Indeed, for each $X \in \Gamma(D)$, $Y \in \Gamma(D^{\perp})$ and $V \in \Gamma(TM^{\perp})$ we have

$g(JX, Y) = -g(X, JY) = 0$, since $JY \in \Gamma(TM^{\perp})$ and

$g(JX, V) = -g(X, JV) = -g(X, BV) = 0$, since $BV \in \Gamma(D^{\perp})$.

Thus M is a CR-submanifold of N. The proof is complete.

Now, suppose M is a CR-submanifold of the almost Hermitian manifold N. Then from (1.5) we obtain

$$\phi^2 = -P \tag{1.9}$$

and

$$\phi^3 + \phi = 0. \tag{1.10}$$

On the other hand, applying J to (1.2) and taking the normal part we get $C^2V + V + \omega BV = 0$, which implies

$$C^3 + C = 0. \tag{1.11}$$

Thus from (1.10) and (1.11) it follows

PROPOSITION 1.1. <u>On each CR-submanifold M the vector bundle morphisms ϕ and C define f-structures on TM and $TM^{\perp}$ respectively.</u>

In order to justify the name CR-submanifold we recall the definition of a CR-manifold. Some results on the geometry of CR-manifolds are found in Chapter VI.

Let M be a differentiable manifold and T_cM be the complexified tangent bundle to M. A CR-structure (see

Greenfield [1]) on M is a complex subbundle $\mathcal{H}$ of $T_C M$ such that $\mathcal{H} \cap \bar{\mathcal{H}} = \{0\}$ and $\mathcal{H}$ is involutive, i.e., for complex vector fields U and V in $\mathcal{H}$, [U, V] is also in $\mathcal{H}$. A manifold endowed with a CR-structure is called a CR-manifold.

THEOREM 1.3 (Blair-Chen [1]). A CR-submanifold of a Hermitian manifold is a CR-manifold.

Proof. Suppose M is a CR-submanifold of the Hermitian manifold N. Then the Nijenhuis tensor of J vanishes. Hence for any X, Y $\in \Gamma(D)$ we have

$$0 = [J, J](X, Y) = -[X, Y] + [JX, JY] - J([JX, Y] + [X, JY]).$$

Thus we obtain

$$\phi([JX, Y] + [X, JY]) = [JX, JY] - [X, Y] \tag{1.12}$$

and

$$Q([JX, JY] - [X, Y]) = 0. \tag{1.13}$$

Applying ϕ to (1.12) and taking account of (1.9) and (1.13) we obtain

$$\phi([X, Y] - [JX, JY]) = [JX, Y] + [X, JY]. \tag{1.14}$$

Now define the complex subbundle $\mathcal{H}$ of $T_C M$ by

$$\mathcal{H}_x = \{X - \sqrt{-1}\,\phi X;\ X \in D_x\}.$$

Take $U = X - \sqrt{-1}\,\phi X$, $V = Y - \sqrt{-1}\,\phi Y$ and by a direct computation we obtain

$$[U, V] = [X, Y] - [JX, JY] - \sqrt{-1}\{[X, JY] + [JX, Y]\}. \tag{1.15}$$

Thus, the theorem follows from (1.15).

§2. Integrability of Distributions on a CR-Submanifold

Let M be a CR-submanifold of an almost Hermitian manifold N. The purpose of this paragraph is to study the integrability of both of the distributions D and $D^\perp$ on M. For each vector field Z tangent to N we denote by Z^T and $Z^\perp$ its tangent part to M and its normal part to M respectively.

The Nijenhuis tensor field of ϕ is given by

$$[\phi, \phi](X, Y) = [\phi X, \phi Y] + \phi^2[X, Y] - \phi([X, \phi Y]) - \phi([\phi X, Y]). \quad (2.1)$$

Then, by using (1.1) and (2.1), we obtain

$$[J, J](X, Y) = [\phi, \phi](X, Y) - Q([X, Y]) - \omega([\phi X, Y] + [X, \phi Y]), \quad (2.2)$$

for any $X, Y \in \Gamma(D)$. Then from (2.2) we have

THEOREM 2.1 (Bejancu [2]). Let M be a CR submanifold of an almost Hermitian manifold N. Then the distribution D is integrable if and only if

$$[J, J](X, Y)^{\top} = [\phi, \phi](X, Y), \quad (2.3)$$

for any $X, Y \in \Gamma(D)$.

Taking the normal part in (2.2) we obtain

$$[J, J](X, Y)^{\perp} = - \omega([\phi X, Y] + [X, \phi Y]), \quad (2.4)$$

for any $X, Y \in \Gamma(D)$.

THEOREM 2.2 (Bejancu [2]). Let M be a CR-submanifold of an almost Hermitian manifold N. Then the distribution D is integrable if and only if

$$[J, J](X, Y)^{\perp} = 0 \quad (2.5)$$

and

$$Q[\phi, \phi](X, Y) = 0, \quad \text{for any } X, Y \in \Gamma(D). \quad (2.6)$$

Proof. Suppose D is integrable. Then (2.5) follows from (2.4). By using (1.9) we obtain

$$[\phi, \phi](X, Y) = [\phi X, \phi Y] - P([X, Y]) - \phi([\phi X, Y] + [X, \phi Y]),$$

for any $X, Y \in \Gamma(D)$. Thus, taking into account that $D = \text{Im}.\phi$ we have

$$[\phi, \phi](X, Y) \in \Gamma(D),$$

which is equivalent to (2.6).

Conversely, suppose (2.5) and (2.6) are satisfied. Then from (2.4) and (2.5) we have

$$Q([JX, Y] + [X, JY]) = 0,$$

which implies

$$Q([JX, JY] - [X, Y]) = 0.$$

Hence $Q([J, J](X, Y)^T) = 0$. On the other hand, from (2.2) we obtain

$$Q([J, J](X, Y)^T) = Q([\phi; \phi](X, Y)) - Q([X, Y]).$$

Thus by (2.6) we obtain $Q([X, Y]) = 0$, that is, D is integrable.

From Theorem 2.1 we obtain

COROLLARY 2.1. <u>Let M be a CR-submanifold of a Hermitian manifold N. The distribution D is integrable if and only if the Nijenhuis tensor of ϕ vanishes identically on D.</u>

Now, we take $X, Y \in \Gamma(D^\perp)$ and obtain

$$[\phi, \phi](X, Y) = -P([X, Y]). \tag{2.7}$$

Thus we have

THEOREM 2.3. <u>Let M be a CR-submanifold of an almost Hermitian manifold N. The distribution $D^\perp$ is integrable if and only if the Nijenhuis tensor of ϕ vanishes identically on $D^\perp$.</u>

Next, we suppose that M is a CR-submanifold of a nearly Kaehlerian manifold N. Then, by using the formulas of Gauss and Weingarten for the immersion of M in N and (5.6) of Chapter I, we obtain

$$\begin{aligned}[JX, Y] + [X, JY] = {} & \frac{1}{2} J([J, J](X, Y)) + J([X, Y]) + \\ & + \nabla_{JX}Y - \nabla_{JY}X + h(JX, Y) - \\ & - h(X, JY),\end{aligned} \tag{2.8}$$

for any $X, Y \in \Gamma(D)$, where ∇ is the Levi-Civita connection on M and h is the second fundamental form of M. Taking into account that ∇ is a torsion-free connection, from (2.8) we get

$$\begin{aligned}h(X, JY) - h(JX, Y) = {} & \frac{1}{2} J([J, J](X, Y)) + \\ & + J([X, Y]) + \nabla_Y JX - \nabla_X JY,\end{aligned} \tag{2.9}$$

for any $X, Y \in \Gamma(D)$.

THEOREM 2.4. (Sato [2]). <u>Let M be a CR-submanifold of a nearly Kaehlerian manifold N. Then the distribution D is integrable if and only if the following conditions are satisfied:</u>

$$h(X, JY) = h(JX, Y). \tag{2.10}$$

and

$$[J, J](X, Y) \in \Gamma(D), \tag{2.11}$$

for any $X, Y \in \Gamma(D)$.

Proof. Suppose D is integrable, Then (2.8) implies

$$h(X, JY) - h(JX, Y) = \frac{1}{2} J([J, J](X, Y)). \tag{2.12}$$

From (2.12), taking account of (2.3) and (2.5) and (2.6), we obtain (2.10). Also, from (2.3) and (2.5) and (2.6) it follows that (2.11) holds.

Conversely, suppose (2.10) and (2.11) are satisfied. Then by using (2.9) we obtain

$$J([X, Y]) = \nabla_X JY - \nabla_Y JX - \frac{1}{2} J([J, J](X, Y)). \tag{2.13}$$

We note that for each $Z \in \Gamma(D^{\perp})$ there exists $V \in \Gamma(TM^{\perp})$ such that $Z = JV$. Then using (2.11) and (2.13) we obtain

$$g([X, Y], JV) = -g(J([X, Y]), V) = 0.$$

Hence $[X, Y] \in \Gamma(D)$ for each $X, Y \in \Gamma(D)$, that is D is integrable.

From Theorem 2.4 and (5.7) of Chapter I we have

THEOREM 2.5 (Urbano [1]). *Let M be a CR-submanifold of a nearly Kaehlerian manifold N. Then the distribution is integrable if and only if*

$$(\tilde{\nabla}_X J)(Y) \in \Gamma(D) \tag{2.14}$$

and (2.10) for any $X, Y \in \Gamma(D)$.

Also, from Theorem 2.5, by again using (5.7) of Chapter I we obtain

COROLLARY 2.2. *Let M be a CR-submanifold of a nearly Kaehlerian manifold N. Then the distribution D is integrable if and only if (2.10) is satisfied and*

$$[J, J](X, U)^{\top} \in \Gamma(D^{\perp}), \tag{2.15}$$

for any $X \in \Gamma(D)$ and $U \in \Gamma(D^{\perp})$.

We denote by ν the complementary orthogonal subbundle to $J(D^{\perp})$ in $TM^{\perp}$. Obviously ν is invariant by J, i.e., $J(\nu_x) = \nu_x$ for each $x \in M$. Then from (2.9) it follows that

$$g(h(X, JY) - h(Y, JX), \xi) =$$

$$= -\frac{1}{2}\{g([J, J](X, Y), J\xi)\} = 0, \quad (2.16)$$

for any $X, Y \in \Gamma(D)$ and $\xi \in \Gamma(\nu)$.

Thus from (2.16) we have

PROPOSITION 2.1. <u>The condition (2.10) is satisfied if and only if</u>

$$g(h(X, JY) - h(Y, JX), JZ) = 0, \quad (2.17)$$

<u>for any $Z, Y \in \Gamma(D)$ and $Z \in \Gamma(D^{\perp})$.</u>

Now we are concerned with integrability of $D^{\perp}$ on a CR-submanifold in a nearly Kaehlerian manifold.

First by using (5.4) and (5.6) of Chapter I we obtain

$$d\Omega(U, W, X) = g((\tilde{\nabla}_U J)W, X), \quad (2.18)$$

for any $U, W \in \Gamma(D^{\perp})$ and $X \in \Gamma(D)$. On the other hand, by direct computation we get

$$3d\Omega(U, W, X) = g([U, W], JX). \quad (2.19)$$

Thus from (2.18) and (2.19) we have

THEOREM 2.6 (Urbano [1]). <u>Let M be a CR-submanifold of a nearly Kaehlerian manifold N. Then the distribution $D^{\perp}$ is integrable if and only if</u>

$$g((\tilde{\nabla}_U J)W, X) = 0, \quad (2.20)$$

<u>for any $U, W \in \Gamma(D^{\perp})$ and $X \in \Gamma(D)$.</u>

Taking account of (5.6) from Chapter I and of the Gauss formula, from Theorem 2.6 we obtain

COROLLARY 2.3 (Sato [2]). <u>Let M be a CR-submanifold of a nearly Kaehlerian manifold N. The distribution $D^{\perp}$ is integrable if and only if</u>

$$g(h(U, X), JW) = g(h(W, X), JU), \quad (2.21)$$

<u>for any $U, W \in \Gamma(D^{\perp})$ and $X \in \Gamma(D)$. Thus we have</u>

PROPOSITION 2.2. <u>Let M be a CR-submanifold of a nearly Kaehlerian manifold N. If $D^{\perp}$ is integrable then each leaf of $D^{\perp}$ is immersed in M as a totally geodesic submanifold if and only if</u>

$$g(h(U, X), JW) = 0, \quad (2.22)$$

for any $U, W \in \Gamma(D^{\perp})$ and $X \in \Gamma(D)$.

By using (5.7) of Chapter I and Theorem 2.6 we obtain

COROLLARY 2.4. Let M be a CR-submanifold of a nearly Kaehlerian manifold N. Then $D^{\perp}$ is integrable if and only if

$$[J, J](X, U)^{T} \in \Gamma(D), \tag{2.23}$$

for any $X \in \Gamma(D)$ and $U \in \Gamma(D^{\perp})$.

Next, combining Corollary 2.2 with Corollary 2.4 we obtain

PROPOSITION 2.3. Let M be a CR-submanifold of a nearly Kaehlerian manifold N. Then both distributions D and $D^{\perp}$ are integrable if and only if (2.10) is satisfied and

$$[J, J](X, U)^{T} = 0, \tag{2.24}$$

for any $X \in \Gamma(D)$ and $U \in \Gamma(D^{\perp})$.

A CR-submanifold of an almost Hermitian manifold is called a mixed geodesic CR-submanifold if its second fundamental form h satisfies

$$h(X, U) = 0, \quad \text{for any} \quad X \in \Gamma(D) \quad \text{and} \quad U \in \Gamma(D^{\perp}). \tag{2.25}$$

Then, by using (3.3) of Chapter I we obtain

PROPOSITION 2.4. Let M be a CR-submanifold of an almost Hermitian manifold N. Then M is mixed geodesic if and only if both distributions D and $D^{\perp}$ are invariant with respect to the action of fundamental tensors of Weingarten, i.e.,

$$A_V X \in \Gamma(D) \quad \text{and} \quad A_V U \in \Gamma(D^{\perp}),$$

for any $X \in \Gamma(D)$, $U \in \Gamma(D^{\perp})$ and $V \in \Gamma(TM^{\perp})$.

Finally, by using (2.25) and Corollary 2.3 we obtain

COROLLARY 2.5. Let M be a mixed geodesic anti-holomorphic submanifold of a nearly Kaehlerian manifold N. Then the distribution $D^{\perp}$ is integrable.

§3. ϕ-Connections on a CR-Submanifold and CR-Products of Almost Hermitian Manifolds

Let M be a CR-submanifold of an almost Hermitian manifold N. In §1 we defined by (1.5) a tensor field ϕ of type (1, 1) on M, which by Proposition 1.1 is an f-structure. A linear connection ∇ on M is called a ϕ-connection if ϕ is covariantly constant with respect to this connection, that is, we have

$$\nabla_X\phi = 0, \quad \text{for each} \quad X \in \Gamma(TM). \tag{3.1}$$

THEOREM 3.1 (Bejancu [10]). All the ϕ-connections on the CR-submanifold M of the almost Hermitian manifold N are given by

$$\nabla_XY = P\overset{\circ}{\nabla}_XPY + Q\overset{\circ}{\nabla}_XQY + \frac{1}{2}\{(\overset{\circ}{\nabla}_X\phi)\phi Y + PK(X, PY) - \phi K(X, \phi Y)\} + QS(X, QY), \tag{3.2}$$

for all X, Y $\in \Gamma(TM)$, where $\overset{\circ}{\nabla}$ is a linear connection with respect to which both distributions D and $D^\perp$ are parallel, P and Q are the projection morphisms to D and $D^\perp$ respectively and K and S are arbitrary tensor fields of type (1, 2) on M.

Remark 3.1. The existence of a linear connection with respect to which both distributions D and $D^\perp$ are parallel is proven by Theorem 4.2 of Chapter I.

Proof of Theorem 3.1. Let ∇ be a ϕ-connection on M. Then put

$$\nabla_XY = \overset{\circ}{\nabla}_XY + S(X, Y), \tag{3.3}$$

for any X, Y $\in \Gamma(TM)$, where $\overset{\circ}{\nabla}$ is a linear connection on M with respect to which both distributions D and $D^\perp$ are parallel and S is a tensor field of type (1, 2) on M. Since ∇ has to satisfy (3.1), by using (3.3) we have

$$(\overset{\circ}{\nabla}_X\phi)Y = \phi S(X, Y) - S(X, \phi Y). \tag{3.4}$$

Substituting ϕY for Y in (3.4) and taking account of (1.9) we obtain

$$(\overset{\circ}{\nabla}_X\phi)Y = \phi S(X, \phi Y) + S(X, PY). \tag{3.5}$$

For each X $\in \Gamma(TM)$ we denote by S_X the tensor field of type (1, 1) on M defined by

$$S_X(Y) = S(X, PY). \tag{3.6}$$

Then (3.5) becomes

$$(\overset{\circ}{\nabla}_X\phi)\circ\phi = S_X + \phi\circ S_X\circ\phi, \tag{3.7}$$

since $P\circ\phi = \phi$. From (3.6) it follows that

$$S_X(Z) = 0, \tag{3.8}$$

for any $Z \in \Gamma(D^{\perp})$ and $X \in \Gamma(TM)$. On the other hand, by using (3.7) and taking into account that D is parallel with respect to $\overset{\circ}{\nabla}$, we obtain

$$Q(S_X(Y)) = 0, \tag{3.9}$$

for any $Y \in \Gamma(D)$ and $X \in \Gamma(TM)$.

Next, we denote by $\Gamma^1_1(TM)$ the real vector space of all tensor fields of type (1, 1) on M. Also, we consider the vector subspace $\Gamma^1_1(D)$ of $\Gamma^1_1(TM)$ defined by

$$\Gamma^1_1(D) = \{H \in \Gamma^1_1(TM);\quad HQY = QHPY = 0,$$
$$\forall Y \in \Gamma(TM)\}. \tag{3.10}$$

Define two linear operators Λ and Ψ on $\Gamma^1_1(D)$ by

$$\Lambda H = \frac{1}{2}(H + \phi\circ H\circ\phi) \tag{3.11}$$

and

$$\Psi H = \frac{1}{2}(H - \phi\circ H\circ\phi). \tag{3.12}$$

By a straightforward computation, using (3.10)-(3.12), we obtain that Λ and Ψ are complementary projectors on $\Gamma^1_1(D)$, that is we have

$$\Lambda^2 = \Lambda;\quad \Psi^2 = \Psi;\quad \Lambda\circ\Psi = \Psi\circ\Lambda = 0;\quad \Lambda + \Psi = 1, \tag{3.13}$$

where I is the identity morphism on $\Gamma^1_1(D)$.

Now, by means of (3.8) and (3.9) we see that $S_X \in \Gamma^1_1(D)$. Hence (3.7) becomes

$$\Lambda(S_X) = \frac{1}{2}(\overset{\circ}{\nabla}_X\phi)\circ\phi. \tag{3.14}$$

We have further that

$$\Psi(\tfrac{1}{2}(\overset{\circ}{\nabla}_X\phi)\circ\phi) = \frac{1}{4}\{(\overset{\circ}{\nabla}_X\phi)\circ\phi + \phi\circ(\overset{\circ}{\nabla}_X\phi)\circ P\} = 0,$$

since $\overset{\circ}{\nabla}_X P = 0$. Thus, there exists S_X such that (3.14) is satisfied and all solutions of (3.14) are given by

$$S_X = \frac{1}{2}\{(\overset{\circ}{\nabla}_X\phi)\circ\phi + K_X - \phi\circ K_X\circ\phi\}, \tag{3.15}$$

where K_X is an arbitrary element of $\Gamma^1_1(D)$. Thus we have

$$S(X, PY) = \frac{1}{2}\{(\overset{\circ}{\nabla}_X\phi)\phi Y + K(X, Y) - \phi K(X, \phi Y)\}, \tag{3.16}$$

where K is a tensor field of type (1, 2) on M satisfying

$$K(X, QY) = QK(X, PY) = 0. \tag{3.17}$$

By using (3.3) we obtain

$$(\nabla_X\phi)QY = -\phi(\overset{\circ}{\nabla}_X QY) - \phi S(X, QY). \tag{3.18}$$

Since ∇ has to be a ϕ-connection and $D^\perp$ is parallel with respect to $\overset{\circ}{\nabla}$, from (3.18) we have

$$PS(X, QY) = 0, \quad \text{for any} \quad X, Y \in \Gamma(TM). \tag{3.19}$$

By using (3.16), (3.17), and (3.19) in (3.3) we obtain (3.2).

Finally, it is a simple verification that all connections connections given by (3.2) are ϕ-connections. This completes the proof of the theorem.

From Theorem 3.1 we obtain

COROLLARY 3.1. Both distributions D and $D^\perp$ on a CR-submanifold of an almost Hermitian manifold are parallel with respect to any ϕ-connection.

Now, we say that a CR-submanifold M of an almost Hermitian manifold N is a CR-product if both distributions D and $D^\perp$ are integrable and M is locally a Riemannian product $M_1 \times M_2$, where M_1 is a leaf of D and M_2 is a leaf of $D^\perp$. A CR-product with $D \neq \{0\}$ and $D^\perp \neq \{0\}$ is called a proper CR-product.

THEOREM 3.2 (Bejancu [10]). Let M be a CR-submanifold of an almost Hermitian manifold N. If the Levi-Civita connection on M is a ϕ-connection, then M is a CR-product.

Proof. By Corollary 3.1 both distributions D and $D^\perp$ are parallel with respect to Levi-Civita connection. Then the assertion of the theorem follows from Theorem 4.4 of Chapter I.

Remark 3.2. Theorem 3.2 has been obtained by Bejancu-

Kon-Yano in [1] for anti-holomorphic submanifolds of a Kaehler manifold and by Chen in [5] for CR-submanifolds of a Kaehler manifold.

THEOREM 3.3 (Sato [2]). Let M be a CR-submanifold of a nearly Kaehlerian manifold N. Suppose the following conditions are satisfied

$$g(h(X, Y), JZ) = 0, \tag{3.20}$$

for any $X \in \Gamma(TM)$, $Y \in \Gamma(D)$, $Z \in \Gamma(D^{\perp})$ and

$$g([J, J](X, Y), W) = 0, \tag{3.21}$$

for any $X, Y \in \Gamma(D)$ and $W \in \Gamma(D^{\perp})$. Then M is a CR-product of N.

Proof. By using (3.3) and (5.6) of Chapter I, (3.20) and the formulas of Gauss and Weingarten for the immersion of M in N we obtain

$$0 = g(\tilde{\nabla}_X Y, JZ) = -g(J\tilde{\nabla}_X Y, Z) =$$
$$= g(J\tilde{\nabla}_Y X - \tilde{\nabla}_X JY - \tilde{\nabla}_Y JX, Z) =$$
$$= -g(h(X, Y), JZ) - g(\nabla_X JY, Z) + g(A_{JX}Y, Z) =$$
$$= -g(\nabla_X JY, Z) + g(h(Y, Z), JX) = -g(\nabla_X JY, Z),$$

for any $Y \in \Gamma(D)$ and $X, Z \in \Gamma(D^{\perp})$. Thus we have

$$\nabla_X JY \in \Gamma(D), \tag{3.22}$$

for any $X \in \Gamma(D^{\perp})$ and $Y \in \Gamma(D)$.

On the other hand, by using (5.7) of Chapter I, (3.20), (3.21) and the formula of Gauss we get

$$0 = g((\tilde{\nabla}_X J)Y, W) = g(\nabla_X JY, W) + g(\nabla_X Y, JW).$$

Hence we have

$$\nabla_X JY \in \Gamma(D), \tag{3.23}$$

for any $X, Y \in \Gamma(D)$. Thus, by (3.22) and (3.23), it follows that D is parallel with respect to the Levi-Civita connection. Our assertion then follows from Proposition 4.1 and Theorem 4.4 of Chapter I.

Remark 3.3. The notion of CR-product in a Kaehler

manifold has been introduced by Chen in [5]. More results on the geometry of CR-products are given in the next paragraph of this chapter, in Chapter III (see §5) and in Chapter IV (see §3).

§4. The Non-Existence of CR-Products in S^6

First we obtain some preliminary results on the geometry of the factors of a CR-product. Then we prove the non-existence of proper CR-products in S^6.

We recall that S^6 is endowed with a natural structure of a nearly Kaehlerian manifold (see §5 of Chapter I).

Let $M = M_1 \times M_2$ be a proper CR-product of S^6. Denote by h' the second fundamental form of the immersion of M_1 in S^6. Then we have

$$h'(X, Y) = h(X, Y), \tag{4.1}$$

for any $X, Y \in \Gamma(TM_1)$, since M_1 is a totally geodesic submanifold of M. Throughout this paragraph we denote by h the second fundamental form of the immersion of M in S^6.

Since M_1 is a holomorphic submanifold of S^6, by a result of Vanhecke [1], it follows that M_1 is a σ-submanifold of S^6, i.e., we have

$$h'(JX, Y) = Jh(X, Y),$$

for any $X, Y \in \Gamma(TM_1)$. Thus by (4.1) we obtain

$$h(JX, Y) = Jh(X, Y). \tag{4.2}$$

Also, by a result due to Gray [1] we have $\dim M_1 = 2$. Next we get

$$g(\nabla_X Y, Z) = -g(Y, \nabla_X Z) = 0, \tag{4.3}$$

for any $X \in \Gamma(TM)$, $Y \in \Gamma(D)$ and $Z \in \Gamma(D^\perp)$ since D and $D^\perp$ are parallel distributions with respect to the Levi-Civita connection on M.

The Gauss formula for the immersion of M in S^6 and (4.3) implies

$$1 + g(h(X, X), h(Z, Z)) - g(h(X, Z), h(X, Z)) = 0, \tag{4.4}$$

$$g(h(X, JX), h(Z, Z)) - g(h(X, Z), h(JX, Z)) = 0, \tag{4.5}$$

$$g(h(X, JX), h(X, Z)) - g(h(X, X), h(JX, Z)) = 0, \tag{4.6}$$

for all unit vector fields $X \in \Gamma(D)$ and $Z \in \Gamma(D^{\perp})$. By using (4.2) we obtain

$$g(h(X, Y), JZ) = -g(h(JX, Y), Z) = 0, \tag{4.7}$$

for any $X, Y \in \Gamma(TM_1)$ and $Z \in \Gamma(TM_2)$,

By means of the Gauss formula and (5.7) from Chapter I we get

$$\tilde{\nabla}_Z JX = \frac{1}{4}\{J([J, J](Z, X))\} + J(\nabla_Z X + h(Z, X)), \tag{4.8}$$

$$\tilde{\nabla}_Z JX = \nabla_Z JX + h(Z, JX). \tag{4.9}$$

Since $\nabla_Z X \in \Gamma(D)$, from (4.8) and (4.9) it follows that

$$h(Z, JX) = \frac{1}{4}\{J([J, J](Z, X))\} + Jh(Z, X). \tag{4.10}$$

Next, we fix a unit vector $Z \in D_x^{\perp}$. Let

$$g(h(E, E), h(Z, Z) = \max_{\substack{\|X\|=1,\\ X\in D_x}} \{g(h(X, X), h(Z, Z))\}. \tag{4.11}$$

Then we have

$$g(h(E, JE), h(Z, Z)) = 0. \tag{4.12}$$

Next, from (4.5) and (4.12) we get

$$g(h(E, Z), h(JE, Z)) = 0. \tag{4.13}$$

LEMMA 4.1. <u>$h(E, Z) \neq 0$ and $\dim M = 3$.</u>

<u>Proof</u>. Suppose $h(E, Z) = 0$. Then from (4.4) it follows that

$$g(h(E, E), h(Z, Z)) + 1 = 0. \tag{4.14}$$

From (4.4) and (4.11) we obtain

$$g(h(E, Z), h(E, Z)) = \max_{\substack{\|X\|=1,\\ X\in D_x}} \{g(h(X, Z), h(X, Z))\}.$$

Then we find $g(h(JE, Z), h(JE, Z)) = 0$, and hence from (4.2) and (4.4) it follows that

$$0 = g(h(JE, JE), h(Z, Z)) + 1 =$$

$$= -g(h(E, E), h(Z, Z)) + 1. \tag{4.15}$$

But (4.15) contradicts (4.14). Thus we have $h(E, Z) \neq 0$. Moreover, we have $\dim M = 3$ since by Proposition 2.2, $h(E, Z)$ is perpendicular to JW for all $W \in D_x^{\perp}$.

Now, from (4.13), taking account of Proposition 2.2 and Lemma 4.1 we obtain

$$h(JE, Z) = aJh(E, Z), \tag{4.16}$$

for some $a \in R$. From (4.2), (4.6) and (4.16) we have

$$\left.\begin{aligned} (a + 1)g(h(E, E), Jh(E, Z)) &= 0, \\ (a + 1)g(h(E, E), h(E, Z)) &= 0. \end{aligned}\right\} \tag{4.17}$$

THEOREM 4.1 (Sekigawa [1]). <u>There does not exist any proper CR-product in S^6.</u>

<u>Proof</u>. Throughout this proof we take $X, Y \in D_x$ and $Z \in D_x^{\perp}$. First, we suppose $a \neq -1$ in (4.17). Then from (4.7) and (4.17) we obtain

$$h(X, Y) = 0. \tag{4.18}$$

From (4.7), taking account of (3.1), (3.2), and (5.7) of Chapter I and of (4.2), (4.3), and (4.18) we obtain

$$g((\nabla_Z h)(X, Y), JZ) = 0, \tag{4.19}$$

where ∇h is given by (3.4) of Chapter I. Thus from Proposition 2.2 and taking account of (3.1), (3.7), and (5.7) of Chapter I, (4.3) and (4.19), we get

$$g(\frac{1}{4} J([J, J](X, Z)) + Jh(X, Z), h(Y, Z)) = 0.$$

Hence by using (5.2) of Chapter I we have

$$\frac{1}{4} g([J, J](JX, Z), h(Y, Z)) = g(Jh(X, Z), h(Y, Z)). \tag{4.20}$$

Now we take $Y = JX$ in (4.20) and obtain

$$\frac{1}{4} g([J, J](JX, Z), h(JX, Z)) =$$

$$= g(Jh(X, Z), h(JX, Z)). \tag{4.21}$$

On the other hand, from (4.10) we have

$$g(Jh(X, Z), h(JX, Z)) = g(h(X, Z), h(X, Z)) -$$

$$- \frac{1}{4} g(h(X, Z), [J, J](X, Z)). \tag{4.22}$$

From (4.22) it follows that

$$g(Jh(X, Z), h(JX, Z)) = g(h(JX, Z), h(JX, Z)) -$$

$$- \frac{1}{4} g(h(JX, Z), [J, J](JX, Z)). \tag{4.23}$$

Then (4.21) and (4.23) imply

$$2g(h(JX, Z), Jh(X, Z)) = g(h(JX, Z), h(JX, Z)). \tag{4.24}$$

From (4.24) it follows that

$$2g(h(JX, Z), Jh(X, Z)) = g(h(X, Z), h(X, Z)). \tag{4.25}$$

Thus by using (4.16), (4.24) and (4.25) we obtain

$$(a^2 - 1)g(h(E, Z), h(E, Z)) = 0$$

and

$$(2a - 1)g(h(E, Z), h(E, Z) = 0.$$

But this is impossible by virtue of Lemma 4.1.
Next, we suppose a = -1 in (4.16). Then from (4.4), taking account of (4.2) and (4.16), we obtain

$$1 + g(h(E, E), h(Z, Z)) - g(h(E, Z), h(E, Z)) = 0,$$

and

$$1 - g(h(E, E), h(Z, Z)) - g(h(E, Z), h(E, Z)) = 0.$$

Thus we have

$$\left.\begin{aligned} g(h(E, E), h(Z, Z)) &= 0, \\ g(h(E, Z), h(E, Z)) &= 1. \end{aligned}\right\} \tag{4.26}$$

By using (4.12), (4.13), (4.16) and (4.26) we get

$$\left.\begin{aligned} g(h(X, Y), h(Z, Z)) &= 0, \\ g(h(X, Z), h(X, Z)) &= 1, \end{aligned}\right\} \tag{4.27}$$

for any unit vectors $X, Y \in D_x$. From (4.7), taking account of (3.1), (3.2) and (5.6) of Chapter I and of (4.2), (4.3) and (4.27) we obtain again (4.19). Thus in a similar way we get a contradiction. The proof is complete.

Remark 4.1. By Theorem 4.1 there exist no proper CR-submanifolds of S^6 with both distributions D and $D^{\perp}$ parallel with respect to Levi-Civita connection. However, Sekigawa constructed in [1] an example of a 3-dimensional proper CR-submanifold such that both distributions D and $D^{\perp}$ are integrable.

Chapter III

CR-SUBMANIFOLDS OF KAEHLERIAN MANIFOLDS

§1. Integrability of Distributions and the Geometry of Leaves

Let N be a Kaehlerian manifold. Then we have

$$(\tilde{\nabla}_X J)Y = 0 \tag{1.1}$$

and consequently

$$[J, J](X, Y) = 0, \tag{1.2}$$

for any $X, Y \in \Gamma(TN)$. Suppose M is a CR-submanifold of N. Then by using (1.1) and (1.2) in Theorems 2.4 and 2.6 of Chapter II we obtain

THEOREM 1.1. *Let M be a CR-submanifold of a Kaehlerian manifold N. Then we have*
(i) *the distribution $D^{\perp}$ is integrable,*
(ii) *the distribution D is integrable if and only if the second fundamental form of M satisfies*

$$h(X, JY) = h(Y, JX), \quad \textit{for any } X, Y \in \Gamma(D). \tag{1.3}$$

Remark 1.1. The assertion (i) is due to Blair-Chen [1] and the assertion (ii) has been obtained by the author in [1]. Also we note that Blair and Chen have constructed in [1] an example of a CR-submanifold of a Hermitian manifold for which $D^{\perp}$ is not integrable.

Remark 1.2. By Proposition 2.1 of Chapter II and the assertion (ii) of Theorem 1.1 we obtain that D is integrable if and only if we have

$$g(h(X, JY) - h(Y, JX), JZ) = 0, \tag{1.4}$$

for any $X, Y \in \Gamma(D)$ and $Z \in \Gamma(D^{\perp})$.

We say that M is a *D-geodesic CR-submanifold* if its second fundamental form satisfies

$$h(X, Y) = 0, \quad \text{for any } X, Y \in \Gamma(D). \tag{1.5}$$

THEOREM 1.2 (Chen [5]). Let M be a CR-submanifold of a Kaehlerian manifold N. Then

(i) the distribution D is integrable and its leaves are totally geodesic in M if and only if

$$g(h(X, Y), JZ) = 0, \tag{1.6}$$

for all $X, Y \in \Gamma(D)$ and $Z \in \Gamma(D^\perp)$,

(ii) the distribution D is integrable and its leaves are totally geodesic in N if and only if M is D-geodesic.

Proof. Suppose D is integrable and each leaf of D is totally geodesic in M. Then we have $\nabla_X Y \in \Gamma(D)$ for any $X, Y \in \Gamma(D)$. By using (1.1) and the formula of Weingarten we obtain

$$g(h(X, Y), JZ) = -g(J(\tilde{\nabla}_X Y), Z) = -g(\tilde{\nabla}_X JY, Z)$$

$$= -g(\nabla_X JY, Z) = 0, \tag{1.7}$$

for any $X, Y \in \Gamma(D)$ and $Z \in \Gamma(D^\perp)$. Thus (1.6) is satisfied.

Conversely, from (1.6) and by Remark 1.2. the integrability of D follows. By a similar computation as in (1.7) we obtain $\nabla_X Y \in \Gamma(D)$ for all $X, Y \in \Gamma(D)$, that is, each leaf of D is totally geodesic in M. Thus we get the assertion (i).

Now, suppose D is integrable and its leaves are totally geodesic in N. Then we have $\tilde{\nabla}_X Y \in \Gamma(D)$ for any $X, Y \in \Gamma(D)$. Thus by using the Gauss formula we obtain

$$g(h(X, Y), V) = g(\tilde{\nabla}_X Y, V) = 0,$$

for any $X, Y \in \Gamma(D)$ and $V \in \Gamma(TM^\perp)$. Hence M is a D-geodesic CR-submanifold.

Finally, suppose (1.5) is satisfied. Then by (1.3) it follows that D is integrable and $\tilde{\nabla}_X Y \in \Gamma(TM)$ for any $X, Y \in \Gamma(D)$. By using (1.1) we have

$$g(\tilde{\nabla}_X Y, Z) = g(\tilde{\nabla}_X JY, JZ) = g(h(X, JY), JZ) = 0,$$

for any $X, Y \in \Gamma(D)$ and $Z \in \Gamma(D^\perp)$. Thus $\tilde{\nabla}_X Y \in \Gamma(D)$, i.e., each leaf of D is totally geodesic in N. The proof is complete.

The invariant distribution D and the anti-invariant distribution $D^\perp$ are defined respectively by the projectors P and Q. Taking account of (1.1) we obtain

$$P(\nabla_X \phi Y) - P(A_{\omega Y} X) = \phi(\nabla_X Y), \tag{1.8}$$

$$Q(\nabla_X \phi Y) - Q(A_{\omega Y} X) = Bh(X, Y), \tag{1.9}$$

$$h(X, \phi Y) + \nabla^{\perp}_X \omega Y = \omega(\nabla_X Y) + Ch(X, Y), \tag{1.10}$$

for any $X, Y \in \Gamma(TM)$, where ϕ and ω are defined by (1.5) and (1.6) respectively of Chapter II.

Differentiating (1.2) of Chapter II and taking account of the decomposition $TN = D \oplus D^{\perp} \oplus TM^{\perp}$ we obtain

$$P(\nabla_X BV) + \phi(A_V X) = P(A_{CV} X), \tag{1.11}$$

$$Q(\nabla_X BV) = Q(A_{CV} X) + B(\nabla^{\perp}_X V), \tag{1.12}$$

$$h(X, BV) + \nabla^{\perp}_X CV + \omega(A_V X) = C(\nabla^{\perp}_X V), \tag{1.13}$$

for all $X \in \Gamma(TM)$ and $V \in \Gamma(TM^{\perp})$.

Now, let M_2 be a leaf of $D^{\perp}$. Then we have

THEOREM 1.3 (Bejancu-Kon-Yano [1]). *A necessary and sufficient condition for the submanifold M_2 to be totally geodesic in M is that*

$$h(X, Z) \in \Gamma(\nu), \tag{1.14}$$

for all $X \in \Gamma(D^{\perp})$, and $Z \in \Gamma(D)$, where ν is the orthogonal complementary subbundle to $JD^{\perp}$ in $TM^{\perp}$.

Proof. We take $X, Y \in \Gamma(D^{\perp})$ in (1.8) and for each $Z \in \Gamma(D)$ we obtain

$$g(\phi(\nabla_X Y), Z) = -g(A_{\omega Y} X, Z) = -g(h(X, Z), \omega Y).$$

This proves our assertion since M_2 is totally geodesic in M if and only if $\nabla_X Y \in \Gamma(D^{\perp})$ for any $X, Y \in \Gamma(D^{\perp})$.

From Theorem 1.3 we have

COROLLARY 1.1. *Let M be a mixed geodesic CR-submanifold of a Kaehlerian manifold N. Then each leaf of $D^{\perp}$ is totally geodesic in M.*

COROLLARY 1.2. *Let M be an anti-holomorphic submanifold of a Kaehlerian manifold N. Then M is mixed geodesic if and only if each leaf of $D^{\perp}$ is totally geodesic in M.*

Concerning the immersion of M_2 in N we have

THEOREM 1.4 (Bejancu [5]). Let M be a CR-submanifold of a Kaehlerian manifold N. Then M_2 is totally geodesic in N if and only if

$$\nabla^{\perp}_X JY \in \Gamma(JD^{\perp}), \quad \text{for any} \quad X, Y \in \Gamma(D^{\perp}) \tag{1.15}$$

and

$$h(X, Z) \in \Gamma(\nu), \quad \text{for any} \quad X \in \Gamma(D^{\perp}) \quad \text{and } Z \in \Gamma(TM). \tag{1.16}$$

Proof. By using (3.1) and (3.2) of Chapter I we obtain

$$g(\tilde{\nabla}_X Y, Z) = g(\tilde{\nabla}_X JY, JZ) = -g(h(X, JZ), JY), \tag{1.17}$$

$$g(\tilde{\nabla}_X Y, JU) = g(h(X, Y), JU) \tag{1.18}$$

and

$$g(\tilde{\nabla}_X Y, V) = g(\tilde{\nabla}_X JY, JV) = g(\nabla^{\perp}_X JY, JV), \tag{1.19}$$

for any $X, Y, U \in \Gamma(D^{\perp})$, $Z \in \Gamma(D)$ and $V \in \Gamma(\nu)$. From (1.17)-(1.19) our assertion follows since M_2 is totally geodesic in N if and only if

$$\tilde{\nabla}_X Y \in \Gamma(D^{\perp}), \quad \text{for all} \quad X, Y \in \Gamma(D^{\perp}).$$

We say that M is $D^{\perp}$-geodesic if

$$h(X, Y) = 0, \quad \text{for all} \quad X, Y \in \Gamma(D^{\perp}). \tag{1.20}$$

THEOREM 1.5 (Chen [5]). Let M be a CR-submanifold of a Kaehlerian manifold N. Then M_2 is totally geodesic in N if and only if M is $D^{\perp}$-geodesic and

$$g(h(X, Z), JY) = 0, \tag{1.21}$$

for any $X, Y \in \Gamma(D^{\perp})$ and $Z \in \Gamma(D)$.

Proof. From (1.19) using (3.1) of Chapter I we obtain

$$g(\nabla^{\perp}_X JY, JV) = g(h(X, Y), V), \tag{1.22}$$

for all $X, Y \in \Gamma(D^{\perp})$ and $V \in \Gamma(\nu)$. Then our assertion follows from Theorem 1.4 using (1.22).

From Theorem 1.5 we have

COROLLARY 1.3. Let M be an anti-holomorphic submanifold of a

Kaehlerian manifold N. Then M_2 is totally geodesic in N if and only if M is both mixed geodesic and $D^\perp$-geodesic.

§2. Umbilical CR-Submanifolds of Kaehlerian Manifolds

First we prove

LEMMA 2.1. Let M be a CR-submanifold of a Kaehlerian manifold N. Then

$$A_{JX}Y = A_{JY}X, \tag{2.1}$$

for all $X, Y \in \Gamma(D^\perp)$.

Proof. By using (3.2) of Chapter I and (1.1) of Chapter II we obtain

$$g(A_{JX}Y - A_{JY}X, Z) = -g([X, Y], \phi Z), \tag{2.2}$$

for any $X, Y \in \Gamma(D^\perp)$ and $Z \in \Gamma(TM)$. Thus our assertion follows from (2.2) taking into account that $D^\perp$ is integrable.

We recall that M is a totally umbilical submanifold if the first and the second fundamental forms are proportional, that is,

$$h(X, Y) = g(X, Y)H, \tag{2.3}$$

for any $X, Y \in \Gamma(TM)$, where H is the mean curvature vector of M. Then we have the following classification theorem for totally umbilical CR-submanifolds of a Kaehlerian manifold.

THEOREM 2.1. (Chen [4] and Bejancu [8]). Let M be a CR-submanifold of a Kaehlerian manifold N. If M is totally umbilical then we have

(i) M is totally geodesic, or
(ii) M is totally real, or
(iii) the anti-invariant distribution $D^\perp$ is one-dimensional.

Proof. Suppose $\dim D^\perp > 1$. From (2.1) we have

$$A_{JX}BH = A_{JBH}X, \tag{2.4}$$

for all $X \in \Gamma(D^\perp)$. Taking account of (2.3) and (2.4) we obtain

$$g(X, X)g(JBH, H) = g(BH, X)g(JX, H). \tag{2.5}$$

By means of (1.2) of Chapter II we see that (2.5) becomes

$$g(X, X)g(BH, BH) = -g(BH, X)g(JX, H). \tag{2.6}$$

We take X orthogonal to BH and from (2.6) obtain BH = 0. Then from (1.11) we obtain

$$P(A_{CH}Y) = \phi(A_H Y), \quad \text{for any} \quad Y \in \Gamma(TM). \tag{2.7}$$

Now suppose M is not totally real, i.e., dim D ⩾ 2. Then take $Z \in \Gamma(D)$ and obtain

$$g(PA_{JH}Y, Z) = g(A_{JH}Y, Z) = g(Y, Z)g(JH, H) = 0 \tag{2.8}$$

and

$$g(\phi(A_H Y), Z) = -g(A_H Y, JZ) = -g(Y, JZ)g(H, H), \tag{2.9}$$

for any $Y \in \Gamma(TM)$. Replacing Y by JZ in (2.9) and taking account of (2.7) and (2.8) we obtain H = 0. Thus M is totally geodesic and the proof is complete.

THEOREM 2.2. Let M be a totally geodesic CR-submanifold of a Kaehlerian manifold N. Then M is a CR-product.

Proof. By using the assertion (i) of Theorem 1.1, Theorem 1.3 and the assertion (i) of Theorem 1.2 we obtain that both distributions D and $D^\perp$ are integrable and their leaves are totally geodesic in M. Thus M is a CR-product.

THEOREM 2.3 (Blair-Chen [1]). Let M be a totally umbilical CR-submanifold of a Kaehlerian manifold N. Then we have

$$K(X \wedge Y) = 0, \tag{2.10}$$

for any $X \in \Gamma(D)$ and $Y \in \Gamma(D^\perp)$, where K is the sectional curvature of N.

Proof. By using (3.4) of Chapter I and (2.3) we obtain

$$(\nabla_X h)(Y, Z) = g(Y, Z)\nabla^\perp_X H, \tag{2.11}$$

for any $X, Y, Z \in \Gamma(TM)$. Equation (3.7) of Chapter I and (2.11) imply

$$\tilde{R}(X, Y, Z, V) = g(Y, Z)g(\nabla^\perp_X H, V) - g(X, Z)g(\nabla^\perp_Y H, V), \tag{2.12}$$

for any $V \in \Gamma(TM^\perp)$. Take $X \in \Gamma(D)$ and $Y \in \Gamma(D^\perp)$. Hence

$JX \in \Gamma(D)$, $JY \in \Gamma(TM^{\perp})$ and from (2.12) obtain

$$\tilde{R}(X, Y, X, Y) = \tilde{R}(X, Y, JX, JY) = 0,$$

which proves our assertion.

COROLLARY 2.1. (i) There exist no proper totally umbilical CR-submanifolds in any positively (or negatively) curved Kaehlerian manifold.

(ii) There exist no totally umbilical real hypersurface of any positively (or negatively) curved Kaehlerian manifold.

Now we prove a pinching theorem for totally umbilical CR-submanifolds of Kaehlerian manifolds.

THEOREM 2.4 (Chen [3]). Let N be a Kaehlerian manifold satisfying the following pinchings for $\delta > 0$,

(a) $0 < \delta \cdot c \leqslant \tilde{H} < c$,

(b) $\tilde{S}(X, X) > \{1 + (1 - \frac{\delta}{2})(s - 1)\}cg(X, X)$,

where $\tilde{H}$ is the holomorphic sectional curvature of N and $\tilde{S}$ is the Ricci tensor of N. Then every totally umbilical CR-submanifold M of N with codimension $\leqslant s$ is a totally geodesic holomorphic submanifold.

Proof. First we suppose that M is of real codimension $2n - m = s$ and $\dim D^{\perp} = s$. In this case we choose $\{E_1, \ldots, E_s\}$ as an orthonormal basis for $D^{\perp}$ and obtain an orthonormal basis $\{JE_1, \ldots, JE_s\}$ of $TM^{\perp}$. Then, by using (2.4) of Chapter I, we obtain

$$\tilde{S}(E_1, E_1) = \tilde{H}(E_1) + \sum_{i=2}^{s} \{\tilde{K}(E_1 \wedge E_i) + \tilde{K}(E_1 \wedge JE_i)\}.$$

By a result of Bishop-Goldberg [1] we have

$$\tilde{S}(E_1, E_1) = \tilde{H}(E_1) + \frac{1}{4}\sum_{i=2}^{s} \{\tilde{H}(E_1+E_i) + \tilde{H}(E_1+JE_i) +$$

$$+ \tilde{H}(E_1-E_i) + \tilde{H}(E_1-JE_i) - \tilde{H}(E_1) - \tilde{H}(E_i)\}.$$

Then, by using (a), we find

$$\tilde{S}(E_1, E_1) < \{1 + (s-1)(1 - \frac{\delta}{2})\}c,$$

which contradicts (b). In a similar way, the theorem follows if $2n - m = s$ and $\dim.D^{\perp} < s$ or $2n - m < s$ and $\dim.D^{\perp} = 2n - m$, or $2n - m < s$ and $\dim.D^{\perp} < 2n - m$.

Therefore we have $\dim.D^{\perp} = 0$ which means M is a holomorphic submanifold. Moreover, it is known (see Ogiue [1]) that any holomorphic submanifold is minimal. Hence M is totally geodesic in N.

Remark 2.1. The assertion of Theorem 2.4 is also valid if we replace (a) and (b) by

(a') $c < \tilde{H} \leqslant \delta c < 0$, and

(b') $\tilde{S}(X, X) < \{1 + (1 - \frac{\delta}{2})(s - 1)\}cg(X, X)$.

Now, let M be a CR-submanifold of a Kaehlerian manifold N. Then the normal bundle to M has the orthogonal decomposition

$$TM^{\perp} = JD^{\perp} \oplus \nu. \tag{2.13}$$

Denote by r the complex dimension of ν_x for each $x \in M$. Since ν is a holomorphic vector bundle we can take a local field of orthonormal frames

$$\{JE_1,\ldots,JE_q, V_1,\ldots,V_r, V_{r+1} = JV_1,\ldots,V_{2r} = JV_r\}$$

on $TM^{\perp}$, where $\{E_1,\ldots,E_q\}$ is a local field of orthonormal frames on $D^{\perp}$. Then we let

$$A_i = A_{JE_i}, \; A_{\alpha} = A_{V_{\alpha}} \quad \text{and} \quad A_{\alpha*} = A_{V_{\alpha*}}.$$

Unless otherwise stated, in this paragraph we use the conventions that the ranges of indices are respectively: $i, j, k, \ldots = 1,\ldots,q$; $\alpha, \beta, \gamma, \ldots = 1,\ldots,r$;

$$\alpha*, \beta*, \gamma*, \ldots = r+1,\ldots,2r.$$

The CR-submanifold M is said to be <u>pseudo-umbilical</u> if the fundamental tensors of Weingarten are given by

$$A_i X = a_i X + b_i g(X, E_i)E_i, \tag{2.14}$$

$$A_{\alpha} X = a_{\alpha} X + \sum_{i=1}^{2} b_{\alpha}^{i} g(X, E_i)E_i, \tag{2.15}$$

and

$$A_{\alpha*}X = a_{\alpha*}X + \sum_{i=1}^{q} b_{\alpha*}^{i} g(X, E_i)E_i, \qquad (2.16)$$

where a_i, b_i, a_α, b_α^i, $a_{\alpha*}$, and $b_{\alpha*}^i$ are differentiable functions on M and $X \in \Gamma(TM)$.

PROPOSITION 2.1. Any pseudo-umbilical CR-submanifold of a Kaehlerian manifold is mixed geodesic.

Proof. From (2.14)-(2.16) it follows that both of the distributions D and $D^\perp$ are invariant under the action of A_V for any $V \in \Gamma(TM^\perp)$. Thus by using Proposition 2.4 of Chapter II we have our assertion.

PROPOSITION 2.2. Let M be a mixed geodesic CR-submanifold of a Kaehlerian manifold N. Then we have

$$A_{JV}X = JA_V X, \qquad (2.17)$$

for any $X \in \Gamma(D)$ and $V \in \Gamma(\nu)$.

Proof. Since M is mixed geodesic we have

$$g(A_{JV}X - JA_V X, Y) = g(h(X, Y), JV) + g(A_V X, JY) = 0,$$

for any $X \in \Gamma(D)$, $Y \in \Gamma(D^\perp)$ and $V \in \Gamma(\nu)$. On the other hand we have

$$g(A_{JV}X - JA_V X, Z) = g(-\tilde{\nabla}_X JV, Z) + g(A_V X, JZ) =$$
$$= g(\tilde{\nabla}_X V + A_V X, JZ) = g(\nabla_X^\perp V, JZ) = 0, \; Z \in \Gamma(D).$$

Thus by using the decomposition (2.13) we obtain (2.17).

PROPOSITION 2.3. Let M be a pseudo-umbilical proper CR-submanifold of a Kaehlerian manifold N. If $q > 1$ then the functions a_i, a_α and $a_{\alpha*}$ vanish identically on M.

Proof. First, by using (2.1) and (2.14) we have

$$a_i = g(A_i E_j, E_j) = g(A_j E_i, E_j) = 0, \quad \text{where} \quad j \neq i.$$

Next, we take a unit vector field X from the invariant distribution D and by using (2.15), (2.16) and Proposition 2.1 we obtain

$$a_{\alpha} = g(A_{\alpha}X, X) = g(JA_{\alpha}X, JX) = g(A_{\alpha*}X, JX) =$$
$$= a_{\alpha*}g(X, JX) = 0.$$

In a similar way we get $a_{\alpha*} = 0$.

THEOREM 2.5 (Bejancu [8]). Let M be a pseudo-umbilical CR-submanifold of a Kaehlerian manifold N. If $q > 1$ then M is a CR-product.

Proof. By using (2.14)-(2.16) we obtain

$$h(X, Y) = \sum_{i=1}^{q} \{b_i g(X, E_i) g(Y, E_i) JE_i\} + \sum_{\alpha=1}^{r} \sum_{i=1}^{q} \{b_{\alpha}^{i} g(X, E_i) g(Y, E_i) V_{\alpha}\} + \sum_{\alpha*=r+1}^{2r} \sum_{i=1}^{q} \{b_{\alpha*}^{i} g(X, E_i) g(Y, E_i) V_{\alpha*}\},$$

for all $X, Y \in \Gamma(TM)$. Thus we have (2.18)

$$h(X, Y) = 0 \quad \text{for any} \quad X \in \Gamma(D) \quad \text{and} \quad Y \in \Gamma(TM).$$

Then by using the assertion (i) of the Theorem 1.2 we obtain that D is integrable and its leaves are totally geodesic in M. On the other hand, by Theorem 1.3 we get that each leaf of $D^{\perp}$ is totally geodesic in M. This completes the proof.

Now, Let M^* be a leaf of $D^{\perp}$. The restrictions of the vector bundles D, $D^{\perp}$ and ν to M^* will be denoted by the same symbols. Thus the normal bundle to M^* is just

$$\mu = D \oplus JD^{\perp} \oplus \nu.$$

For each $V \in \Gamma(\nu)$ we put

$$JV = B^*V + C^*V, \tag{2.19}$$

where $B^*V \in \Gamma(D^{\perp})$ and $C^*V \in \Gamma(D \oplus \nu)$. Since M^* is a totally real submanifold of N, the morphism C^* is an f-structure on the normal bundle μ (see Yano-Kon [1]). Denote by h^* the second fundamental form of M^* in N and by A_V^* the fundamental tensor of Weingarten corresponding to the normal section V. For amy $V \in \Gamma(\mu)$ and $X \in \Gamma(TM^*)$ we define the covariant derivative of C^* by

$$(\nabla'_X C^*)V = \nabla^*_X C^*V - C^*(\nabla^*_X V),$$

where ∇^* is the linear connection induced by $\tilde{\nabla}$ on the vector bundle μ. Yano and Kon obtained in [1]

$$(\nabla'_X C^*)V = -h^*(X, B^*V) - J(A^*_V X), \tag{2.20}$$

for any $X \in \Gamma(TM^*)$ and $V \in \Gamma(\mu)$. The f-structure C^* is said to be <u>parallel</u> if we have $\nabla'_X C^* = 0$ for any $X \in \Gamma(TM^*)$. Also, the f-structure C on $TM^\perp$ defined in §1 of Chapter II is said to be <u>parallel</u> if we have

$$(\nabla_X C)V = \nabla^\perp_X CV - C(\nabla^\perp_X V) = 0,$$

for any $X \in \Gamma(TM)$ and $V \in \Gamma(TM^\perp)$.

THEOREM 2.6. <u>Let M be a pseudo-umbilical proper CR-submanifold of a Kaehlerian manifold N with $q > 1$. Then the following assertions are equivalent to each other</u>

(i) <u>the f-structure C is parallel on $TM^\perp$,</u>
(ii) <u>the f-structure C^* is parallel on μ,</u>
(iii) <u>all the functions b^1_α and $b^1_{\alpha*}$ vanish identically on M.</u>

<u>Proof</u>. From (1.13) we obtain

$$(\nabla_X C)V = -h(X, BV) - \omega(A_V X), \tag{2.21}$$

for any $X \in \Gamma(TM)$ and $V \in \Gamma(TM^\perp)$. Replacing V by JE_i and then by V_α and $V_{\alpha*}$ in (2.21) and taking account of (2.18) we obtain the equivalence of assertions (i) and (iii).

Note that Proposition 2.1 and Corollary 1.1 imply that M^* is totally geodesic in M. Hence, for any $X, Y \in \Gamma(TM^*)$, we have

$$h^*(X, Y) = h(X, Y). \tag{2.22}$$

From (2.22) it follows that $A^*_V = 0$ for any $V \in \Gamma(D)$. Thus, by (2.20) we obtain that C^* is parallel if and only if

$$h^*(X, E_i) = J(A^*_{JE_i} X) \tag{2.23}$$

and

$$A^*_{V_\alpha} X = A^*_{V_{\alpha*}} X = 0, \tag{2.24}$$

for all $X \in \Gamma(TM^*)$.

On the other hand, from (2.18) and (2.21) it follows that $\nabla_X C = 0$ whenever $X \in \Gamma(D)$. Hence C is parallel on $TM^{\perp}$ if and only if

$$h(X, E_i) = J(A_i X) \tag{2.25}$$

and

$$A_{\alpha} X = A_{\alpha *} X = 0, \tag{2.26}$$

for any $X \in \Gamma(D^{\perp})$.

By using (2.22)-(2.26) we conclude that C^* is parallel if and only if C is parallel. This completes the proof.

Taking account of (3.4) from Chapter I and (2.22) we obtain

PROPOSITION 2.4. Let M be a pseudo-umbilical proper CR-submanifold of a Kaehlerian manifold N with $q > 1$ and parallel second fundamental form. Then each leaf of $D^{\perp}$ has parallel second fundamental form.

Now we state a result on the existence of pseudo-umbilical proper CR-submanifolds in Kaehlerian manifolds.

THEOREM 2.7. There exist no pseudo-umbilical proper CR-submanifolds with $q > 1$ immersed in a positively (or negatively) curved Kaehlerian manifold.

Proof. By using (3.7) of Chapter I and (2.18) we obtain

$$g(\tilde{R}(X, Y)JX, JY) = -g(h(Y, \nabla_X JX), JY), \tag{2.27}$$

for any $X \in \Gamma(D)$ and $Y \in \Gamma(D^{\perp})$. Since $h(X, JX) = h(X, X) = 0$, from (3.1) of Chapter I it follows that

$$\nabla_X JX = J(\nabla_X X). \tag{2.28}$$

Hence $\nabla_X JX \in \Gamma(D)$ and by using (2.27) and Proposition 2.1 we have

$$g(\tilde{R}(X, Y)X, Y) = g(\tilde{R}(X, Y)JX, JY) = 0.$$

Thus, taking account of (2.6) from Chapter I, we obtain the assertion of the theorem.

§3. Normal CR-Submanifolds of Kaehlerian Manifolds

Let M be a CR-submanifold of a Kaehlerian manifold N. Then

by (1.5) and (1.6) of Chapter II we have an f-structure ϕ on M and a normal bundle-valued 1-form ω on M. Taking into account that g is a Hermitian metric on N we get

$$g(\phi X, \phi Y) = g(X, Y) - g(\omega X, \omega Y), \tag{3.1}$$

for any X, Y $\in \Gamma(TM)$. Also the covariant derivative of ϕ is defined by

$$(\nabla_X \phi)Y = \nabla_X(\phi Y) - \phi(\nabla_X Y), \tag{3.2}$$

for any X, Y $\in \Gamma(TM)$. Then by using (1.8) and (1.9) we obtain

$$(\nabla_X \phi)Y = Bh(X, Y) + A_{\omega Y}X.$$

On the other hand the covariant derivative of ω is defined by

$$(\nabla_X \omega)Y = \nabla^{\perp}_X(\omega Y) - \omega(\nabla_X Y), \tag{3.3}$$

for any X, Y $\in \Gamma(TM)$. By (1.10) and (3.3) we have

$$(\nabla_X \omega)Y = Ch(X, Y) - h(X, \phi Y). \tag{3.4}$$

Also by using (1.11)-(1.13) we obtain

$$(\nabla_X B)V = A_{CV}X - \phi(A_V X) + B(\nabla^{\perp}_X V) \tag{3.5}$$

and

$$(\nabla_X C)V = -h(X, BV) - \omega(A_V X), \tag{3.6}$$

for any X $\in \Gamma(TM)$ and V $\in \Gamma(TM^{\perp})$.

The fundamental 2-form Ω of a CR-submanifold M in a Kaehlerian manifold N is defined by

$$\Omega(X, Y) = g(X, \phi Y), \tag{3.7}$$

for any X, Y $\in \Gamma(TM)$. By using (1.9) of Chapter II and (3.7) we obtain

$$\Omega(\phi X, \phi Y) = \Omega(X, Y). \tag{3.8}$$

In fact Ω is nothing but the restriction of the fundamental 2-form of the Kaehlerian manifold N to TM $\times$ TM. Hence Ω is closed, i.e., we have $d\Omega = 0$.

The exterior derivative of ω is given by

$$d\omega(X, Y) = \frac{1}{2}\{\nabla^{\perp}_X(\omega Y) - \nabla^{\perp}_Y(\omega X) - \omega([X, Y])\}. \tag{3.9}$$

Now we state

PROPOSITION 3.1. <u>Let M be a CR-submanifold of a Kaehlerian</u>

manifold N. Then the covariant derivative of ϕ is given by

$$2g((\nabla_X\phi)Y, Z) = g([\phi, \phi](Y, Z), \phi X) + 2g(d\omega(\phi Y, X), \omega Z) + 2g(d\omega(\phi Y, Z) - d\omega(\phi Z, Y), \omega X) - 2g(d\omega(\phi Z, X), \omega Y), \quad (3.10)$$

for all vector fields X, Y, Z tangent to M.

Proof. Since the Riemannian connection of M is given by

$$2g(\nabla_X Y, Z) = Xg(Y, Z) + Yg(X, Z) - Zg(X, Y) + g([X, Y], Z) + g([Z, X], Y) - g([Y, Z], X),$$

we have

$$\begin{aligned} 2g((\nabla_X\phi)Y, Z) &= 2g(\nabla_X\phi Y, Z) + 2g(\nabla_X Y, \phi Z) = \\ &= Xg(\phi Y, Z) + \phi Yg(X, Z) - Zg(X, \phi Y) + \\ &+ g([X, \phi Y], Z) + g([Z, X], \phi Y) - g([\phi Y, Z], X) + \\ &+ Xg(Y, \phi Z) + Yg(X, \phi Z) - \phi Zg(X, Y) + \\ &+ g([X, Y], \phi Z) + g([\phi Z, X], Y) - g([Y, \phi Z], X). \end{aligned}$$

Thus, taking into account that g is covariantly constant with respect to ∇ and by using (2.1) of Chapter II we obtain

$$\begin{aligned} 2g((\nabla_X\phi)Y, Z) &= g([\phi, \phi](Y, Z), \phi X) + \\ &+ g(\nabla^{\perp}_{\phi Y}\omega Z - \omega([\phi Y, Z]), \omega X) + \\ &+ g(\nabla^{\perp}_{\phi Y}\omega X - \omega([\phi Y, X]), \omega Z) + g(\omega X, \omega([\phi Z, Y])) + \\ &+ g(\omega Y, \omega([\phi Z, X])) - \phi Zg(\omega X, \omega Y). \end{aligned} \quad (3.11)$$

Next, from (3.9) it follows that

$$d\omega(\phi Y, Z) = \frac{1}{2}\{\nabla^{\perp}_{\phi Y}\omega Z - \omega([\phi Y, Z])\}. \quad (3.12)$$

Since $\nabla^{\perp}$ is a Riemannian connection on the normal bundle we have

$$\phi Zg(\omega X, \omega Y) = g(\nabla^{\perp}_{\phi Z}\omega X, \omega Y) + g(\omega X, \nabla^{\perp}_{\phi Z}\omega Y). \quad (3.13)$$

Finally, by using (3.12) and (3.13) in (3.11) we obtain (3.10).

We define the tensor field S by

$$S(X, Y) = [\phi, \phi](X, Y) - 2Bd\omega(X, Y), \tag{3.14}$$

for any $X, Y \in \Gamma(TM)$. Substituting $d\omega$ from (3.9) and taking into account that ∇ is a torsion-free linear connection, (3.14) becomes

$$S(X, Y) = (\nabla_{\phi X}\phi)Y - (\nabla_{\phi Y}\phi)X + \phi\{(\nabla_Y\phi)X - (\nabla_X\phi)Y\} - \\ - B\{(\nabla_X\omega)Y - (\nabla_Y\omega)X\}. \tag{3.15}$$

The CR-submanifold M is said to be <u>normal</u> if the tensor field S vanishes identically on M. If, in particular, M is a real hypersurface of a Kaehlerian manifold we get the well known notion of a normal real hypersurface of a Kaehlerian manifold (see Okumura [1], [3]).

THEOREM 3.1. (Bejancu [7]). <u>The CR-submanifold M is normal if and only if we have</u>

$$A_{\omega Y}(\phi X) = \phi(A_{\omega Y}X), \tag{3.16}$$

<u>for all $X \in \Gamma(D)$ and $Y \in \Gamma(D^\perp)$.</u>

<u>Proof</u>. By using (3.2) and (3.4) in (3.15) we obtain

$$S(X, Y) = (A_{\omega Y}\circ\phi - \phi\circ A_{\omega Y})X - (A_{\omega X}\circ\phi - \phi\circ A_{\omega X})Y, \tag{3.17}$$

for all $X, Y \in \Gamma(TM)$.

Suppose M is a normal CR-submanifold of N. Then (3.16) follows from (3.17) since $A_{\omega X} = 0$ for any $X \in \Gamma(D)$. Now, if (3.16) is satisfied we shall prove $S = 0$ by means of the decomposition $TM = D \oplus D^\perp$. First, from (3.17) we have that $S(X, Y) = 0$ for all $X, Y \in \Gamma(D)$. Next, for $X \in \Gamma(D)$ and $Y \in \Gamma(D^\perp)$ we obtain

$$S(X, Y) = (A_{\omega Y}\circ\phi - \phi\circ A_{\omega Y})X = 0,$$

by (3.16). Finally, for $X, Y \in \Gamma(D^\perp)$, (3.17) becomes

$$S(X, Y) = \phi(A_{\omega X}Y - A_{\omega Y}X),$$

since $\phi X = \phi Y = 0$. By using (2.1) we obtain the vanishing of S on $D^\perp \times D^\perp$. The proof is complete.

Now, suppose $\{E_1,\ldots,E_q\}$ is a local field of orthonormal frames for the anti-invariant distribution $D^\perp$. Denote by A_i the fundamental tensor of Weingarten with respect to $V_i = JE_i$ $(i = 1,\ldots,q)$. Then from Theorem 3.1 we have

COROLLARY 3.1. The CR-submanifold M is normal if and only if the fundamental tensors of Weingarten A_i commute with ϕ on the invariant distribution, that is we have

$$A_i \circ \phi = \phi \circ A_i \quad (i = 1,\ldots,q). \tag{3.18}$$

Remark 3.1. Corollary 3.1 is a generalization of a result due to Okumura [3] for normal real hypersurfaces of a Kaehlerian manifold.

Using (3.1) and (3.2) of Chapter I for the immersion of M in N we obtain

$$\nabla_X E_i = \phi(A_i X) - B(\nabla^\perp_X V_i) \tag{3.19}$$

and

$$\nabla^\perp_X V_i = \omega(\nabla_X E_i) + Ch(X, E_i), \quad (i = 1,\ldots,q), \tag{3.20}$$

for all $X \in \Gamma(TM)$.

It is well known that X is a Killing vector field if and only if we have

$$g(\nabla_Z X, Y) + g(Z, \nabla_Y X) = 0,$$

for any $Y, Z \in \Gamma(TM)$. We introduce here a weaker condition for X. Thus we say that X is a D-Killing vector field if we have

$$g(\nabla_Z X, Y) + g(Z, \nabla_Y X) = 0, \tag{3.21}$$

for any $Y, Z \in \Gamma(D)$.

Now we state

THEOREM 3.2. A necessary and sufficient condition for the CR-submanifold to be normal is that E_i $(i = 1,\ldots,q)$ be D-Killing vector fields.

Proof. Using (3.19) we obtain

$$g(\nabla_Z E_i, Y) + g(Z, \nabla_Y E_i) = g((\phi \circ A_i - A_i \circ \phi)Z, Y), \tag{3.22}$$

for all $Y, Z \in \Gamma(D)$. Thus our assertion follows from (3.22) and Corollary 3.1.

The Lie derivative of ϕ with respect to $Y \in \Gamma(TM)$ is given by

$$(\mathcal{L}_Y\phi)X = [Y, \phi X] - \phi([Y, X]), \tag{3.23}$$

for any $X \in \Gamma(TM)$. Then normal CR-submanifolds can be characterized by another tensor field S^* defined by

$$S^*(Y, X) = (\mathcal{L}_Y\phi)X, \tag{3.24}$$

for any $X, Y \in \Gamma(TM)$.

THEOREM 3.3. <u>Suppose M is a CR-submanifold of a Kaehlerian manifold N and we have</u>

$$Q(\nabla_X Y) = 0, \tag{3.25}$$

<u>for all $X \in \Gamma(D)$ and $Y \in \Gamma(D^\perp)$. Then M is a normal CR-submanifold of N if and only if</u>

$$S^*(Y, X) = 0, \tag{3.26}$$

<u>for all $X \in \Gamma(D)$ and $Y \in \Gamma(D^\perp)$.</u>

<u>Proof</u>. From the proof of Theorem 3.1 follows that M is a normal CR-submanifold if and only if $S(X, Y) = 0$ for any $X \in \Gamma(D)$ and $Y \in \Gamma(D^\perp)$. By using (3.4) and (3.9) in (3.14) we obtain

$$S(X, Y) = \phi(\phi([X, Y]) - [\phi X, Y]) - Bh(\phi X, Y). \tag{3.27}$$

Next, from (1.10) we have

$$h(\phi X, Y) = \omega(\nabla_Y X) + Ch(X, Y),$$

which implies

$$Bh(\phi X, Y) = -Q(\nabla_Y X).$$

Hence (3.27) becomes

$$S(X, Y) = \phi(\phi([X, Y]) - [\phi X, Y]) \pm Q(\nabla_Y X). \tag{3.28}$$

Using (3.23), (3.24), and (3.28) we obtain

$$S(X, Y) = \phi(S^*(Y, X)) + Q(\nabla_Y X). \tag{3.29}$$

Now suppose M is a normal CR-submanifold. Then from (3.29) we obtain

$$PS^*(Y, X) = 0 \text{ and} \tag{3.30}$$

$$Q(\nabla_Y X) = 0. \tag{3.31}$$

Also, by using (3.23), (3.24), (3.25), and (3.31), we have

$QS^*(Y, X) = 0$. Therefore (3.26) is satisfied.

Conversely, suppose (3.26) is satisfied. Then from (3.23) and (3.24) it follows that

$$Q([X, Y]) = 0, \tag{3.32}$$

for any $X \in \Gamma(D)$ and $Y \in \Gamma(D^\perp)$. Thus, by using (3.25), (3.26) and (3.32) in (3.29), we obtain $S(X, Y) = 0$. Thus M is a normal CR-submanifold.

By using Corollary 3.1 it is easy to verify that totally geodesic CR-submanifolds, totally umbilical CR-submanifolds and pseudo-umbilical CR-submanifolds are examples of normal CR-submanifolds. Moreover, each normal real hypersurface of a Kaehlerian manifold is also an example of normal CR-submanifold.

§4. Normal Anti-Holomorphic Submanifolds of Kaehlerian Manifolds

In this paragraph we consider an anti-holomorphic submanifold M of a Kaehlerian manifold N. Thus the dimension of the anti-invariant distribution on M is the same as the codimension of the immersion of M in N. We introduce some tensor fields which are generalizations of tensor fields introduced by Sasaki and Hatakeyama in [1] for an almost contact structure on a manifold. We shall obtain characterizations of normal anti-holomorphic submanifolds by means of these tensor fields and Ricci curvatures.

Suppose $\dim D^\perp = q$ and choose a local field of orthonormal frames $\{E_1,\dots,E_q\}$ on the anti-invariant distribution $D^\perp$. Then define the following tensor fields

$$S^{(1)}(X, Y) = [\phi, \phi](X, Y) - 2Jd\omega(X, Y), \tag{4.1}$$

$$S^{(2)}(X, Y) = 2J\{d\omega(\phi X, Y) - d\omega(\phi Y, X)\}, \tag{4.2}$$

$$S^{(3)}(x) = (\mathcal{L}_{E_i}\phi)(X), \quad (i = 1,\dots,q), \tag{4.3}$$

for any X, Y tangent to M.

Remark 4.1. The tensor field $S^{(1)}$ is just the tensor field S defined on each CR-submanifold by (3.14). Also, we remark that the tensor fields $S_i^{(3)}$ depend on the local field of orthonormal frames $\{E_1,\dots,E_q\}$. However, in this section, since we are dealing only with anti-holomorphic submanifolds, we prove their independence of this field of frames.

Since M is an anti-holomorphic submanifold, from (3.2) and (3.4) we obtain

$$(\nabla_X\phi)Y = Jh(X, Y) + A_{\omega Y}X \tag{4.4}$$

and

$$(\nabla_X\omega)Y = -h(X, \phi Y), \tag{4.5}$$

for any $X, Y \in \Gamma(TM)$.

PROPOSITION 4.1. Let M be a mixed geodesic anti-holomorphic submanifolds of a Kaehlerian manifold N. Then $S^{(2)}$ vanishes identically on M.

Proof. By using (3.9) and (4.5) in (4.2) we obtain

$$S^{(2)}(X, Y) = J\{h(QX, PY) - h(QY, PX)\}.$$

Thus our assertion follows from (2.25) of Chapter II.

COROLLARY 4.1. Let M be a mixed geodesic anti-holomorphic submanifold of a Kaehlerian manifold N. Then $d\omega$ is invariant by ϕ, that is, we have

$$d\omega(\phi X, \phi Y) = d\omega(X, Y), \tag{4.6}$$

for any $X, Y \in \Gamma(TM)$.

Proof. Since $S^{(2)} = 0$, by (4.2) we have

$$d\omega(\phi X, \phi Y) = d\omega(X, PY), \tag{4.7}$$

for any $X, Y \in \Gamma(TM)$. On the other hand, by using (3.9) and (4.5), we obtain

$$d\omega(X, QY) = \frac{1}{2}\{h(QY, \phi X) - h(X, \phi QY)\} = 0. \tag{4.8}$$

Thus (4.6) follows from (4.7) and (4.8).

PROPOSITION 4.2. Each normal anti-holomorphic submanifold of a Kaehlerian manifold is mixed geodesic.

Proof. By using (3.3) of Chapter I we have

$$g(h(E_j, X), JE_i) = g(A_iX, E_j),\ i, j = 1,\dots,q,$$

for any $X \in \Gamma(D)$. By Theorem 1.1 of Chapter II we have $\mathrm{Im}\ \phi = D$. Hence there exist $Y \in \Gamma(TM)$ such that $X = \phi Y$.

Thus, by using (3.18), we obtain

$$g(h(X, E_j), JE_i) = g(\phi A_i Y, E_j) = g(A_i \phi Y, E_j) = 0.$$

This proves our assertion.

In the remaining part of this section we are dealing with anti-holomorphic submanifolds with flat normal connection, i.e., the curvature tensor $R^{\perp}$ of $\nabla^{\perp}$ vanishes identically on M.

PROPOSITION 4.3. Let M be a normal anti-holomorphic submanifold of a Kaehlerian manifold N with flat normal connection. Then we have

(i) $S^{(2)}$ vanishes identically on M,

(ii) $S_i^{(3)}$ vanishes identically on M for any local field of orthonormal frames on $D^{\perp}$ and i = 1,...,q.

Proof. By Propositions 4.1 and 4.2 we have the assertion (i). In order to prove the assertion (ii) we choose a distinguished field of frames on the anti-invariant distribution. By Theorem 3.1 of Chapter I there exists a local field of orthonormal frames $\{V_1,\ldots,V_q\}$ such that V_i is parallel with respect to the normal connection. Then the distinguished field of frames on $D^{\perp}$ is defined by

$$\{E_1 = -JV_1,\ldots,E_q = -JV_q\}. \tag{4.9}$$

By using (4.3) the tensor fields $S_i^{(3)}$ corresponding to the field of frames (4.9) are given by

$$S_i^{(3)}(X) = (\nabla_{E_i}\phi)X - (\nabla_X\phi)E_i - \nabla_{\phi X}E_i. \tag{4.10}$$

Now take $X \in \Gamma(D)$ and by (4.4) and Proposition 4.2 obtain

$$(\nabla_{E_i}\phi)X = Jh(X, E_i) = 0 \quad \text{and} \quad (\nabla_X\phi)E_i = A_iX.$$

Thus, by using (3.18), (3.19), and (4.10), we have

$$S_i^{(3)}(X) = -A_iX - (\phi\circ A_i\circ\phi)X = 0.$$

The tensor fields $S_i^{(3)}$ vanish on the anti-invariant distribution if and only if $S_i^{(3)}(E_j) = 0$ for any E_j from the distinguished field of frames (4.9). By (3.19) and (4.3) we obtain

$$S_i^{(3)}(E_j) = -\phi([E_i, E_j]) = -\phi(\nabla_{E_i} E_j - \nabla_{E_j} E_i) =$$
$$= -\phi(\phi A_j E_i - \phi A_i E_j) = 0.$$

Hence the tensor fields $S_i^{(3)}$ vanish identically on M.

Next, suppose $\{\bar{E}_i\}$ $(i = 1,\ldots,q)$ is an arbitrary field of orthonormal frames on $D^{\perp}$. Denote by $\bar{S}_i^{(3)}$ the corresponding tensor fields given by (4.3). First we claim that $\bar{S}_i^{(3)}(X) \in \Gamma(D)$ for each $X \in \Gamma(TM)$. In fact, this follows from the equalities

$$g(\nabla_{\bar{E}_i} \phi X, E_j) = -g(\phi X, \nabla_{\bar{E}_i} E_j) = -g(\phi X, \phi A_j \bar{E}_i) = 0$$

and

$$g(\nabla_{\phi X} \bar{E}_i, E_j) = -g(\bar{E}_i, \nabla_{\phi X} E_j) = -g(\bar{E}_i, \phi A_j \phi X) = 0,$$

by means of (4.3). Finally we put

$$\bar{E}_i = \sum_{j=1}^{q} f_i^j E_j.$$

Then we have

$$\bar{S}_i^{(3)}(X) = \sum_{j=1}^{q} \{f_i^j S_j^{(3)}(X) - \phi X(f_i^j) E_j\} = 0,$$

since $S_j^{(3)}(X) = 0$ and $\bar{S}_i^{(3)}(X) \in \Gamma(D)$. The proof is complete.

THEOREM 4.1. <u>Let M be an anti-holomorphic submanifold of a Kaehlerian manifold N with flat normal connection. Then the following assertions are equivalent to each other:</u>

(i) <u>M is a normal anti-holomorphic submanifold of N;</u>

(ii) <u>the vector fields E_i from the distinguished field of frames (4.9) are Killing vector fields;</u>

(iii) <u>the tensor fields $S_i^{(3)}$ vanish identically on M.</u>

<u>Proof</u>. By (3.19), taking account of flatness of the normal connection we obtain

$$g(\nabla_Z E_i, Y) + g(\nabla_Y E_i, Z) = g((\phi \circ A_i - A_i \circ \phi)Y, Z), \tag{4.11}$$

for all vector fields Y, Z tangent to M. Thus (4.11) proves the equivalence of assertions (i) and (ii) by means of Corollary 3.1.

Next, by Proposition 4.3, we see that (i) implies (iii). Suppose $S_i^{(3)} = 0$. Using (3.19) and (4.4) in (4.3) we obtain

$$0 = S_i^{(3)}(X) = Jh(E_i, X) + \phi(\phi A_i X - A_i \phi X),$$

for any $X \in \Gamma(D)$. Thus by Corollary 3.1 M is a normal anti-holomorphic submanifold. This completes the proof.

Now combining Theorems 3.2 and 4.1 we obtain

COROLLARY 4.2. Let M be an anti-holomorphic submanifold of a Kaehlerian manifold N with flat normal connection. Then the vector fields E_i from the distinguished field of frames are Killing vector fields if and only if they are D-Killing vector fields.

Let M be a mixed geodesic anti-holomorphic submanifold of a Kaehlerian manifold N. Then by Proposition 2.4 of Chapter II we have the invariance of the holomorphic distribution by the action of fundamental tensors of Weingarten $A_i = A_{JE_i}$. Denote by $A_{i|D}$ the restriction of A_i to D.

THEOREM 4.2 (Bejancu [14]). Let M be a mixed geodesic anti-holomorphic submanifold of a Kaehlerian manifold N with flat normal connection. Then M is a normal anti-holomorphic submanifold if and only if the Ricci curvatures with respect to the distinguished field of frames (4.9) are given by

$$k(E_i) = \text{trace}\,(A_{i|D})^2, \quad i = 1,\ldots,q. \tag{4.12}$$

Proof. First, by using (3.19), we obtain the curvature tensor R of M. It is given by

$$R(E_i, X)E_i = \nabla_{E_i} \phi A_i X - \phi A_i([E_i, X]), \tag{4.13}$$

for all X tangent to M. For any $X \in \Gamma(D)$, by (3.19) and Proposition 2.4 of Chapter II we have

$$g(\nabla_{E_i} X, E_j) = -g(X, \nabla_{E_i} E_j) = 0, \quad i, j = 1,\ldots,q.$$

Hence $\nabla_{E_i} X \in \Gamma(D)$. Then replacing X by E_i and Y by PZ in (4.4) we obtain

$$\nabla_{E_i} \phi Z = J(\nabla_{E_i} PZ), \tag{4.14}$$

for any $Z \in \Gamma(TM)$. We take $X \in \Gamma(D)$ and by (4.13) and (4.14) obtain

$$g(R(E_i, X)E_i, X) = g(J(\nabla_{E_i} A_i X), X) - g(JA_i(\nabla_{E_i} X), X) - g(JA_i X, A_i JX). \tag{4.15}$$

Thus we have

$$\begin{aligned} g(R(E_i, X)E_i, X) &+ g(R(E_i, JX)E_i, JX) = \\ &= g(J(\nabla_{E_i} A_i X), X) - g(JA_i(\nabla_{E_i} X), X) + \\ &+ g(\nabla_{E_i} A_i JX, X) - g(A_i(\nabla_{E_i} X), X) - \\ &- 2g(JA_i X, A_i JX) = g(\nabla_{E_i}(JA_i + A_i J)X, X) + \\ &+ g((JA_i + A_i J)X, \nabla_{E_i} X) - 2g(JA_i X, A_i JX). \end{aligned} \tag{4.16}$$

By direct computation it follows that

$$g((JA_i + A_i J)X, X) = 0.$$

Hence, (4.16) becomes

$$g(R(E_i, X)E_i, X) + g(R(E_i, JX)E_i, JX) = = -2g(JA_i X, A_i JX) \tag{4.17}$$

for any $X \in \Gamma(D)$.

Now, for any $X \in \Gamma(D^{\perp})$ from (4.13) it follows that

$$R(E_i, X)E_i = 0. \tag{4.18}$$

since $D^{\perp}$ is integrable and invariant by A_i.

Further, we choose the following local field of orthonormal frames on M

$$\{F_1, \ldots, F_p, F_1^* = JF_1, \ldots, F_p^* = JF_p, E_1, \ldots, E_q\}, \tag{4.19}$$

where $F_a \in \Gamma(D)$, $a = 1, \ldots, p$, $\dim {}_C D_x = p$ and $\{E_1, \ldots, E_q\}$ is the distinguished field of frames (4.9) on $D^{\perp}$. Then by using (4.17)-(4.19), the Ricci curvature $k(E_i)$ becomes

$$k(E_i) = 2 \sum_{a=1}^{p} \{g(JA_i F_a, A_i JF_a)\}. \tag{4.20}$$

By a straightforward computation we obtain

$$g((A_iJ - JA_i)X, (A_iJ - JA_i)X) = g(A_iX, A_iX) + $$
$$+ g(A_iJX, A_iJX) - 2g(A_iJX, JA_iX), \tag{4.21}$$

for all $X \in \Gamma(D)$.

Next, taking into account that A_i are symmetric linear operators and g is a Hermitian metric we obtain that $A_iJ - JA_i$ are symmetric linear operators too. Denote by $(A_iJ - JA_i)_{|D}$ the restrictions of $A_iJ - JA_i$ to D. Then by using (4.21) and (4.20) we obtain

$$k(E_i) = \text{trace } (A_{i|D})^2 - \frac{1}{2} \text{ trace } (A_iJ - JA_i)^2_{|D}. \tag{4.22}$$

Now suppose M is a normal anti-holomorphic submanifold. Then by using (3.18) and (4.22) we obtain (4.12).

Conversely, if (4.12) is satisfied, by (4.22) we have

$$\text{trace } (A_iJ - JA_i)^2_{|D} = 0. \tag{4.23}$$

All the eigenvalues of $A_iJ - JA_i$ are real since they are symmetric linear operators. Hence the eigenvalues of $(A_iJ - JA_i)^2_{|D}$ are non-negative. Thus (4.23) implies

$$(A_iJ - JA_i)_{|D} = 0.$$

Therefore, by Corollary 3.1 M is a normal anti-holomorphic submanifold. The proof is complete.

THEOREM 4.3. Let M be a mixed geodesic anti-holomorphic submanifold of a Kaehlerian manifold N with flat normal connection. Then each leaf of the anti-invariant distribution $D^\perp$ is a locally Euclidean space.

Proof. Suppose M^* is a leaf of $D^\perp$. Denote by h^* the second fundamental form of the immersion of M^* in M and by R^* the curvature tensor of M^*. Then the Gauss equation with respect to the immersion of M^* in M becomes

$$g(R(X, Y)Z, W) = g(R^*(X, Y)Z, W) + $$
$$+ g(h^*(X, Z), h^*(Y, W)) - g(h^*(Y, Z), h^*(X, W)), \tag{4.24}$$

for all X, Y, Z, W $\in \Gamma(TM^*)$. By using (3.19) we obtain $R(X, Y)Z = 0$, for any X, Y, $Z \in \Gamma(D^\perp)$. On the other hand, by Corollary 1.2 we obtain the vanishing of h^* on M^*. Then from (4.24) it follows that $R^* = 0$, which means M^* is a locally Euclidean space.

Combining Theorem 4.3 with Proposition 4.2 we have

COROLLARY 4.3. Let M be a normal anti-holomorphic submanifold of a Kaehlerian manifold with flat normal connection. Then each leaf of the anti-invariant distribution is a locally Euclidean space.

§5. CR-Products in Kaehlerian Manifolds

By Theorem 3.2 of Chapter II we found a class of CR-submanifolds which are CR-products in almost Hermitian manifolds. We show here that if the ambient space is a Kaehlerian manifold this is the only class of CR-products. More precisely we have

PROPOSITION 5.1. Let M be a CR-product in a Kaehlerian manifold N. Then

$$\nabla\phi = 0, \tag{5.1}$$

that is, the Levi-Civita connection is a ϕ-connection.

Proof. First, for any $U \in \Gamma(TM)$ and $Y \in \Gamma(D)$ we have $\nabla_U Y \in \Gamma(D)$. By using (3.1) of Chapter I and (1.2) we obtain

$$g(h(U, JY), V) = g(\tilde{\nabla}_Y JY, V) = g(J(\tilde{\nabla}_U Y), V) =$$

$$= g(Jh(U, Y), V),$$

for any $V \in \Gamma(TM^\perp)$. Hence $Bh(U, Y) = 0$ and by (3.2) we have $(\nabla_U\phi)Y = 0$. Finally, for $X \in \Gamma(D^\perp)$ and $U \in \Gamma(TM)$ we get $\nabla_U X \in \Gamma(D^\perp)$. Hence $(\nabla_U\phi)X = 0$ and we obtain our assertion.

THEOREM 5.1 (Chen [5]). Let M be a CR-submanifold of a Kaehlerian manifold N. Then M is a CR-product if and only if (5.1) is satisfied.

The proof follows by combining Theorem 3.2 of Chapter II and Proposition 5.1.

THEOREM 5.2. Let M be a CR-submanifold of a Kaehlerian manifold N. Then M is a CR-product if and only if

$$Bh(X, Y) = 0, \tag{5.2}$$

for any $X \in \Gamma(TM)$ and $Y \in \Gamma(D)$.

Proof. Suppose M is a CR-product. Then by using (3.2) and Theorem 5.1 we obtain (5.2).

Conversely, suppose (5.2) be satisfied. Then from (3.2) it follows that

$$(\nabla_X \phi) Y = 0 \quad \text{for any} \quad X \in \Gamma(TM) \quad \text{and} \quad Y \in \Gamma(D).$$

Now take $Z \in \Gamma(D^{\perp})$ and by using (1.8) obtain

$$(\nabla_X \phi) Z = -\phi(\nabla_X Z) = P(A_{JZ} X),$$

for any $X \in \Gamma(TM)$. Finally, (3.3) of Chapter I and (5.2) imply

$$g(A_{JZ} X, Y) = g(h(X, Y), JZ) = -g(Bh(X, Y), Z) = 0,$$

for any $Y \in \Gamma(D)$. Thus we have $(\nabla_X \phi) Z = 0$. Hence by Theorem 3.2 of Chapter II M is a CR-product.

From Theorem 5.2 it follows that

COROLLARY 5.1. Let M be an anti-holomorphic submanifold of a Kaehlerian manifold N. Then M is a CR-product if and only if

$$h(X, Y) = 0, \tag{5.3}$$

for any $X \in \Gamma(TM)$ and $Y \in \Gamma(D)$.

Now we state

THEOREM 5.3. Each CR-product in a Kaehlerian manifold is a normal CR-submanifold.

Proof. Let M be a CR-product in a Kaehlerian manifold N. Then by Theorem 5.2 we have

$$g(h(X, Y), \omega Z) = 0, \tag{5.4}$$

for any $X \in \Gamma(TM)$, $Y \in \Gamma(D)$ and $Z \in \Gamma(D^{\perp})$. Thus by using (5.4) we obtain

$$g(A_{\omega Z}\phi Y - \phi(A_{\omega Z} Y), X) = g(h(X, \phi Y), \omega Z) +$$

$$+ g(h(Y, \phi X), \omega Z) = 0.$$

Hence by Theorem 3.1 M is a normal CR-submanifold.

LEMMA 5.1. Let M be a CR-product in a Kaehlerian manifold N. Then for any unit vector fields $X \in \Gamma(D)$ and $Y \in \Gamma(D^{\perp})$ we have

$$\tilde{H}_B(X \wedge Y) = 2\|h(X, Y)\|^2, \tag{5.5}$$

where $\tilde{H}_B(X \wedge Y)$ is the holomorphic bisectional curvature of N determined by $\{X, Y\}$.

Proof. By using (3.7) of Chapter I, (5.2) and taking into account that D is a parallel distribution we obtain

$$\tilde{R}(X, JX, Y, JY) = g(\nabla^{\perp}_X h(JX, Y) - \nabla^{\perp}_{JX} h(X, Y), JY). \tag{5.6}$$

Next, by using (5.4) we see that (5.6) becomes

$$\begin{aligned}\tilde{R}(X, JX, Y, JY) &= g(h(X, Y), \nabla^{\perp}_{JX} JY) - \\ &- g(h(JX, Y), \nabla^{\perp}_X JY) = g(h(X, Y), Jh(JX, Y)) - \\ &- g(h(JX, Y), Jh(X, Y)) = 2g(h(X, Y), Jh(JX, Y)).\end{aligned} \tag{5.7}$$

for any $X \in \Gamma(D)$ and $Y \in \Gamma(D^{\perp})$. By using (3.3) of Chapter I and (5.2) we obtain

$$h(JX, Y) = Jh(X, Y). \tag{5.8}$$

Then (5.5) follows from (5.7) and (5.8) and (5.12) in Chapter I.

From Lemma 5.1 we have

THEOREM 5.4 (Chen [5]). Let N be a Kaehlerian manifold with negative holomorpnic bisectional curvature. Then every CR-product in N is either a holomorphic submanifold or a totally real submanifold.

We say that a CR-product M is an anti-holomorphic CR-product if the dimension of a fibre of the anti-invariant distribution is just the codimension of M in N. Then we have

THEOREM 5.5. Let M be an anti-holomorphic CR-product of a Kaehlerian manifold N with positive or negative holomorphic bisectional curvature. Then M is a totally real submanifold.

Proof. Since M is an anti-holomorphic CR-product each leaf of the anti-invariant distribution is totally geodesic in M. By Corollary 1.2 M is a mixed geodesic CR-submanifold. Hence by (5.5) we have

$$\tilde{H}_B(X \wedge Y) = 0, \quad \text{for any} \quad X \in \Gamma(D) \quad \text{and} \quad Y \in \Gamma(D^{\perp}),$$

from which our assertion is proved.

We say that the anti-holomorphic submanifold M is a <u>cosymplectic anti-holomorphic submanifold</u> if it is normal and the differential form ω is closed, that is, we have

$$d\omega = 0. \tag{5.9}$$

Also we say that ω and ϕ are parallel on M if we have $\nabla\omega = 0$ and $\nabla\phi = 0$ respectively.

PROPOSITION 5.2. <u>Let M be an anti-holomorphic submanifold of a Kaehlerian manifold N. Then the f-structure ϕ is parallel if and only if the differential form ω is parallel.</u>

<u>Proof</u>. By Theorem 5.1 ϕ is parallel if and only if M is a CR-product. On the other hand, from (4.5) we see that ω is parallel if and only if (5.3) is satisfied. Thus the assertion follows by using Corollary 5.1.

THEOREM 5.6 (Bejancu [7]). <u>Let M be an anti-holomorphic submanifold of a Kaehlerian manifold N. Then M is a cosymplectic submanifold if and only if M is a CR-product.</u>

<u>Proof</u>. Suppose M is a cosymplectic anti-holomorphic submanifold of N. Since M is normal and $d\omega = 0$, from (3.14) we obtain $[\phi, \phi] = 0$. Thus by (3.10) we have $\nabla\phi = 0$, that is, by Theorem 5.1 M is a CR-product.

Conversely, suppose M is a CR-product. Then by Propositions 5.1 and 5.2 we have $\nabla\phi = 0$ and $\nabla\omega = 0$. Hence $d\omega = 0$. On the other hand, from (3.14) we have $S = 0$. Thus M is a normal anti-holomorphic submanifold.

THEOREM 5.7. <u>An anti-holomorphic submanifold M of a Kaehlerian manifold N is cosymplectic if and only if we have</u>

$$(\nabla_X\phi)X = 0, \tag{5.10}$$

<u>for any X tangent to M.</u>

<u>Proof</u>. Suppose M is a cosymplectic anti-holomorphic submanifold of N. Then (5.10) is satisfied since, by Theorems 5.6 and 5.1, ϕ is parallel on M.

Conversely, suppose (5.10) is satisfied on M. Then replacing X and Y from (4.4) by X + Y and taking account of (5.10) we obtain

$$2Jh(X, Y) + A_{\omega Y}X + A_{\omega X}Y = 0, \tag{5.11}$$

for any X, Y $\in \Gamma(TM)$. If we take X, Y $\in \Gamma(D)$ then (5.11) implies

$$h(X, Y) = 0. \tag{5.12}$$

Next, for X $\in \Gamma(D^{\perp})$ and Y $\in \Gamma(D)$, (5.11) becomes

$$2Jh(X, Y) + A_{\omega X}Y = 0. \tag{5.13}$$

On the other hand, from (1.9) we obtain

$$Jh(X, Y) + Q(A_{\omega X}Y) = 0. \tag{5.14}$$

Thus from (5.13) and (5.14), taking account that $Jh(X, Y) \in \Gamma(D^{\perp})$ we obtain

$$h(X, Y) = 0, \tag{5.15}$$

for any X $\in \Gamma(D^{\perp})$ and Y $\in \Gamma(D)$. Finally, we see that (5.12) and (5.15) imply (5.3). Thus M is a CR-product and by Theorem 5.6 a cosymplectic manifold. The proof is complete.

§6. Sasakian Anti-Holomorphic Submanifolds of Kaehlerian Manifolds

Let M be a proper anti-holomorphic submanifold of a Kaehlerian manifold N. Take the local field of orthonormal frames on D given by

$$\{F_1,\ldots,F_p, JF_1,\ldots,JF_p\}. \tag{6.1}$$

Define a normal vector field H_D by

$$H_D = \frac{1}{2p}\sum_{i=1}^{p}\{h(F_i, F_i) + h(JF_i, JF_i)\}. \tag{6.2}$$

Thus H_D is a well-defined normal vector field to M and it is called the D-mean curvature vector of M.

We say that M is a contact anti-holomorphic submanifold of N if $H_D \neq 0$ and we have

$$d\omega(X, Y) = -\Omega(X, Y)H_D, \tag{6.3}$$

for all vector fields X, Y tangent to M. By using (3.9) and (4.5) we see that (6.3) is equivalent to

$$h(\phi X, Y) - h(\phi Y, X) = -2\Omega(X, Y)H_D. \tag{6.4}$$

PROPOSITION 6.1. Let M be a contact anti-holomorphic submanifold of a Kaehlerian manifold N. Then M is mixed geodesic and D is not involutive.

Proof. We take $X \in \Gamma(D)$ and $Y \in \Gamma(D^\perp)$ in (6.4) and obtain that M is mixed geodesic. Take a non-null vector field $X \in \Gamma(D)$ and $Y = JX$ in (6.4). Then we obtain

$$h(JX, JX) + h(X, X) = 2g(X, X)H_D \neq 0,$$

that is (1.3) is not satisfied. Thus D is not involutive.

A Sasakian anti-holomorphic submanifold is a normal contact anti-holomorphic submanifold of N. Any Sasakian real hypersurface of a Kaehlerian manifold is an example of a Sasakian anti-holomorphic submanifold.

Suppose M is an anti-holomorphic submanifold with flat normal connection. Then denote by $\{E_1,\ldots,E_q\}$ the distinguished field of frames (4.9) on $D^\perp$ induced by parallel unit normal sections $\{V_1,\ldots,V_q\}$. Thus (3.19) becomes

$$\nabla_X E_i = \phi(A_i X), \quad (i = 1,\ldots,q), \tag{6.5}$$

for any X tangent to M.

THEOREM 6.1. Let M be an anti-holomorphic submanifold of a Kaehlerian manifold N with flat normal connection. Then the following assertions are equivalent to each other:

(i) M is a Sasakian anti-holomorphic submanifold;

(ii) the covariant derivative of each vector field from the distinguished field of frames (4.9) is given by

$$\nabla_X E_i = g(H_D, V_i)\phi X, \tag{6.6}$$

for any $X \in \Gamma(TM)$ and $i = 1,\ldots,q$.

(iii) the fundamental tensors of Weingarten satisfy

$$A_i(\phi X) = g(H_D, V_i)\phi X, \tag{6.7}$$

for any $X \in \Gamma(TM)$ and $i = 1,\ldots,q$.

Proof. Suppose M is a Sasakian anti-holomorphic submanifold of N. Then by using (3.3) of Chapter I, (3.18), (4.5) and (6.5) we obtain

$$d\omega(X, Y) = \sum_{i=1}^{q} \{g(\nabla_X E_i, Y)V_i\}, \tag{6.8}$$

for any $X, Y \in \Gamma(TM)$.

On the other hand, since M is a contact anti-holomorphic submanifold we have

$$d\omega(X, Y) = g(\phi X, Y) \sum_{i=1}^{q} \{g(H_D, V_i)V_i\}. \tag{6.9}$$

Thus (6.6) follows from (6.8) and (6.9).

Now suppose (6.6) holds. Then for all $Y, Z \in \Gamma(TM)$ we have

$$g(\nabla_Y E_i, Z) + g(Y, \nabla_Z E_i) =$$

$$= g(H_D, V_i)\{g(\phi Y, Z) + g(\phi Z, Y)\} = 0.$$

Hence by Theorem 4.1 M is a normal anti-holomorphic submanifold. We note that in order to obtain (6.8) we made use only of the fact that M is a normal anti-holomorphic submanifold. Thus, by using (6.5) in (6.8), we obtain (6.3), that is, M is a contact anti-holomorphic submanifold. Hence the equivalence of assertions (i) and (ii) is proven.

Further, we suppose (ii) is satisfied. Then from (6.5) and (6.6) it follows that

$$\phi(A_i X) = g(H_D, V_i)\phi X. \tag{6.10}$$

On the other hand, we have seen that (6.6) implies the normality of M. Hence from (6.10) and (3.18) it follows that (6.7) holds.

Conversely, suppose (6.7) is satisfied. Then by Proposition 2.4 of Chapter II M is a mixed geodesic anti-holomorphic submanifold. If we take $X \in \Gamma(D^\perp)$ then from (6.5) it follows that $\nabla_X E_i = 0$. Hence (6.7) implies (6.6) in this case. For any $X \in \Gamma(D)$ there exists $Y \in \Gamma(D)$ such that $\phi Y = X$. Then by (6.7) we have

$$\phi(A_i X) = \phi(A_i \phi Y) = g(H_D, V_i)\phi^2 Y = g(H_D, V_i)\phi X. \tag{6.11}$$

From (6.11) and (6.5) it follows that (6.6) holds for each $X \in \Gamma(D)$. The proof is complete.

THEOREM 6.2. <u>Let M be an anti-holomorphic submanifold of a Kaehlerian manifold N with flat normal connection. Then M</u>

is a Sasakian anti-holomorphic submanifold if and only if

$$(\nabla_X \phi)Y = g(\phi X, \phi Y)JH_D + g(\omega Y, H_D)PX, \qquad (6.12)$$

for any $X, Y \in \Gamma(TM)$.

Proof. Suppose (6.12) is satisfied and replacing Y by E_i we obtain

$$\phi(\nabla_X E_i) = -g(H_D, V_i)PX.$$

Thus by (6.5) we have (6.6). Hence, by Theorem 6.1, M is a Sasakian anti-holomorphic submanifold of N.

Conversely, suppose M is a Sasakian anti-holomorphic submanifold of N. Then from (4.1) we have

$$g([\phi, \phi](Y, Z), \phi X) = 0,$$

for any $X, Y, Z \in \Gamma(TM)$. Thus, by (3.10) and (6.3), we have

$$g((\nabla_X \phi)Y, Z) = g(\phi Z, \phi X)g(H_D, \omega Y) -$$

$$- g(\phi Y, \phi X)g(H_D, \omega Z) = g(Z, g(H_D, \omega Y)PX) +$$

$$+ g(\phi X, \phi Y)g(JH_D, Z) = g(g(H_D, \omega Y)PX +$$

$$+ g(\phi X, \phi Y)JH_D, Z).$$

Hence we have (6.12) and the proof is complete.

Now let Ω be the fundamental 2-form of M defined by (3.7). Then, by using (1.2) of Chapter I, we obtain

$$(\nabla_X \Omega)(Y, Z) = g((\nabla_X \phi)Y, Z), \qquad (6.13)$$

for any $X, Y, Z \in \Gamma(TM)$. Thus from (6.12) and (6.13) we obtain

THEOREM 6.3. Let M be an anti-holomorphic submanifold of a Kaehlerian manifold N with flat normal connection. Then M is a Sasakian anti-holomorphic submanifold if and only if

$$(\nabla_X \Omega)(Y, Z) = g(\phi X, \phi Z)g(H_D, \omega Y) -$$

$$- g(\phi X, \phi Y)g(H_D, \omega Z), \qquad (6.14)$$

for any $X, Y, Z \in \Gamma(TM)$.

Now we state

THEOREM 6.4. Let M be an anti-holomorphic submanifold of a

Kaehlerian manifold N with flat normal connection. Then M is a Sasakian anti-holomorphic submanifold if and only if there exist q(q+1)/2 differentiable sections L_{ij} $(L_{ij} = L_{ji})$ of the normal bundle such that

$$h(X, Y) = g(\phi X, \phi Y)H_D + \sum_{i,j=1}^{q} \{g(Y, E_j)g(\omega X, L_{ij})V_i\} \tag{6.15}$$

for any X, Y $\in \Gamma(TM)$.

Proof. By direct computation we obtain

$$(\tilde{\nabla}_X\Omega)(Y, Z) = (\nabla_X\Omega)(Y, Z) - \Omega(h(X, Y), Z) - - \Omega(Y, h(X, Z)).$$

Since N is Kaehlerian we have

$$(\tilde{\nabla}_X\Omega)(Y, Z) = g((\tilde{\nabla}_X J)Y, Z) = 0.$$

Hence we get

$$(\nabla_X\Omega)(Y, Z) = g(h(X, Z), JY) - g(h(X, Y), JZ), \tag{6.16}$$

for any X, Y, Z $\in \Gamma(TM)$.

Now, suppose M is a Sasakian anti-holomorphic submanifold. Then from (6.14) and (6.16) it follows that

$$g(h(X, Z), JY) - g(h(X, Y), JZ) = = g(\phi X, \phi Z)g(H_D, \omega Y) - g(\phi X, \phi Y)g(H_D, \omega Z). \tag{6.17}$$

On the other hand, by (2.1) we have

$$g(h(X, Z), JY) - g(h(X, Y), JZ) = 0, \tag{6.18}$$

for any X $\in \Gamma(TM)$ and Y, Z $\in \Gamma(D^{\perp})$. Next, we take $Z = E_i$ in (6.17) and taking account of (6.18) and of Proposition 6.1 we obtain

$$g(h(X, Y), JE_i) = g(\phi X, \phi Y)g(H_D, V_i) + + g(h(QY, E_i), \omega X).$$

Thus we have

$$h(X, Y) = g(\phi X, \phi Y)H_D + \sum_{i=1}^{q} \{g(h(QY, E_i), \omega X)V_i\}. \tag{6.19}$$

Substituting

$$QY = \sum_{j=1}^{q} \{Y^j E_j\}$$

in (6.19) we obtain

$$h(X, Y) = g(\phi X, \phi Y)H_D + \sum_{i,j=1}^{q} \{Y^j g(h(E_j, E_i), \omega X)V_i\},$$

which implies (6.15) with $L_{ij} = h(E_i, E_j)$.

Conversely, suppose (6.15) is satisfied. Then (6.14) follows from (6.16) since we have

$$\sum_{i,j=1}^{q} \{(g(Y, E_j)g(V_i, JZ) - g(Z, E_j)g(V_i, JY))g(L_{ij}, \omega X)\} = 0.$$

Hence by Theorem 6.3 M is a Sasakian anti-holomorphic submanifold. The proof is complete.

Now, from (6.7) it follows that on each Sasakian anti-holomorphic submanifold with flat normal connection we have

$$A_i X = g(H_D, V_i)X, \tag{6.20}$$

for any $X \in \Gamma(D)$ and $i = 1,\ldots,q$. Thus, by using (4.12) and (6.20) we obtain

PROPOSITION 6.2. Let M be a Sasakian anti-holomorphic submanifold of a Kaehlerian manifold N with flat normal connection. Then the Ricci curvatures of M with respect to the distinguished field of frames (4.9) are given by

$$k(E_i) = 2pg(H_D, V_i)^2. \tag{6.21}$$

With respect to the sectional curvatures of M we have

PROPOSITION 6.3. Let M be a Sasakian anti-holomorphic submanifold of a Kaehlerian manifold N with flat normal connection. Then we have

$$K_M(E_i \wedge X) = g(H_D, V_i)^2 \tag{6.22}$$

and

$$K_M(E_i \wedge E_j) = 0, \tag{6.23}$$

for any unit vector field $X \in \Gamma(D)$ and $i, j = 1,\ldots,q$, $i \neq j$.

Proof. By using (6.6) we obtain

$$R(E_i, X)E_i = \nabla_{E_i}(g(H_D, V_i)\phi X) -$$
$$- g(H_D, V_i)\phi([E_i, X]) = E_i(g(H_D, V_i))\phi X +$$
$$+ g(H_D, V_i)(\nabla_{E_i}\phi)X - g(H_D, V_i)^2 X, \tag{6.24}$$

for any unit vector field $X \in \Gamma(D)$. But from (4.4) it follows that $(\nabla_{E_i}\phi)X = 0$, since M is mixed geodesic. By using (2.6) of Chapter I and (6.24) we have

$$K_M(E_i \wedge X) = g(R(E_i, X)X, E_i) = -g(R(E_i, X)E_i, X) =$$
$$= g(H_D, V_i)^2.$$

Finally, by (6.6) we have $\nabla_{E_i}E_j = 0$ which implies $R(E_i, E_j)E_i = 0$. This completes the proof.

From Proposition 6.3 we have

COROLLARY 6.1. <u>There exist no positively or negatively curved Sasakian anti-holomorphic submanifolds with flat normal connection and $q > 1$ in any Kaehlerian manifold.</u>

§7. Cohomology of CR-Submanifolds

First we give some remarks on distributions on Riemannian manifolds (see §4 of Chapter I).

Let $\mathcal{H}$ be a differentiable distribution on a Riemannian manifold M with Levi-Civita connection ∇. We put

$$\overset{\circ}{h}(X, Y) = (\nabla_X Y)^{\perp}, \tag{7.1}$$

for any $X, Y \in \Gamma(\mathcal{H})$, where $(\nabla_X Y)^{\perp}$ denotes the component of $\nabla_X Y$ in the orthonormal complementary distribution $\mathcal{H}^{\perp}$ on M. Let $\{X_1,\dots,X_r\}$ be an orthonormal basis of $\mathcal{H}$, $r = \dim._R \mathcal{H}$. We put

$$\overset{\circ}{H} = \frac{1}{r}\sum_{i=1}^{r} \overset{\circ}{h}(X_i, X_i). \tag{7.2}$$

Then $\overset{\circ}{H}$ is a well-defined $\mathcal{H}^{\perp}$-valued vector field on M, called the <u>mean curvature vector</u> of $\mathcal{H}$. The distribution $\mathcal{H}$ is called <u>minimal</u> if the mean curvature vector $\overset{\circ}{H}$ of $\mathcal{H}$ vanishes identically.

Let M be a CR-submanifold of a Kaehlerian manifold N. Then we have

LEMMA 7.1. <u>The holomorphic distribution D on M is minimal.</u>

<u>Proof</u>. By using the formulas of Gauss and Weingarten for the immersion of M in N we obtain

$$g(Z, \nabla_X X) = g(JZ, \tilde{\nabla}_X JX) = -g(\tilde{\nabla}_X JZ, JX) =$$

$$= g(A_{JZ}X, JX), \tag{7.3}$$

for any $X \in \Gamma(D)$ and $Z \in \Gamma(D^\perp)$. Then we have

$$g(Z, \nabla_{JX}JX) = -g(A_{JZ}JX, X) = -g(A_{JZ}X, JX). \tag{7.4}$$

Thus by (7.3) and (7.4) we obtain

$$g(Z, \nabla_X X + \nabla_{JX}JX) = 0,$$

which proves our assertion.

Now, we choose an orthonormal local field of frames $\{F_1,\ldots,F_p, JF_1,\ldots,JF_p\}$ of D. We denote by $\{\omega^1,\ldots,\omega^p, \omega^{p+1},\ldots,\omega^{2p}\}$ the 1-forms on M satisfying

$$\omega^j(Z) = 0,\ \omega^i(F_j) = \delta^i_j, \quad i, j = 1,\ldots,2p, \tag{7.5}$$

for any $Z \in \Gamma(D^\perp)$ where $F_{p+j} = JF_j$. Then

$$\omega = \omega^1 \wedge \ldots \wedge \omega^{2p} \tag{7.6}$$

is a globally well-defined 2p-form on M because D is orientable.

THEOREM 7.1 (Chen [7]). <u>Let M be a closed CR-submanifold of a Kaehlerian manifold N. Then ω is closed and thus defines a canonical de Rham cohomology class given by</u>

$$c(M) = [\omega] \in H^{2p}(M; R), \quad p = \dim._C D. \tag{7.7}$$

<u>Moreover, this cohomology class is nontrivial if D is integrable and $D^\perp$ is minimal.</u>

<u>Proof</u>. From (7.6) we obtain

$$d\omega = \sum_{i=1}^{2p} (-1)^i \omega^1 \wedge \ldots \wedge d\omega^i \wedge \ldots \wedge \omega^{2p} \tag{7.8}$$

By using (7.5) and (7.8) we obtain that $d\omega = 0$ if and only if

$$d\omega(Z_1, Z_2, X_1, \ldots, X_{2p-1}) = 0 \tag{7.9}$$

and

$$d\omega(Z_1, X_1, \ldots, X_{2p}) = 0, \tag{7.10}$$

for any $Z_1, Z_2 \in \Gamma(D^\perp)$ and $X_1, \ldots, X_{2p} \in \Gamma(D)$. By a straightforward computation it follows that (7.9) holds if and only if $D^\perp$ is integrable and (7.10) holds if and only if D is minimal. But by Theorem 1.1 and Lemma 7.1 $D^\perp$ is integrable and D is minimal. Hence the 2p-form ω is closed.

Now, let $\{F_{2p+1}, \ldots, F_{2p+q}\}$ be an orthonormal local field of frames of $D^\perp$ and let $\{\omega^{2p+1}, \ldots, \omega^{2p+1}\}$ be the 1-forms on M satisfying

$$\omega^\alpha(X) = 0 \quad \text{and} \quad \omega^\alpha(F_\beta) = \delta^\alpha_\beta,$$

for any $X \in \Gamma(D)$ and $\alpha, \beta = 2p+1, \ldots, 2p+q$. Then in a similar way we may conclude that if D is integrable and $D^\perp$ is minimal, then the q-form $\omega^\perp = \omega^{2p+1} \wedge \ldots \wedge \omega^{2p+q}$ is closed. Thus the 2p-form ω is coclosed, i.e. $\delta\omega = 0$. Since M is a closed submanifold, ω is harmonic. Because ω is nontrivial, the cohomology class $[\omega]$ represented by ω is nontrivial in $H^{2p}(M; R)$. Thus the proof is complete.

Next we choose an orthonormal local field of frames

$$\{F_1, \ldots, F_p, F_{p+1}, \ldots, F_{p+q}, F_{p+q+1}, \ldots, F_n,$$
$$, JF_1, \ldots, JF_n\}$$

on N in such a way that restricted to M, $\{F_1, \ldots, F_p, JF_1, \ldots, JF_p\}$ are in D and $\{F_{p+1}, \ldots, F_{p+q}\}$ are in $D^\perp$. We denote by $\{\omega^1, \ldots, \omega^n, \omega^{1*}, \ldots, \omega^{n*}\}$ the dual frame to $\{F_1, \ldots, F_n, JF_1, \ldots, JF_n\}$. We put

$$\theta^A = \omega^A + \sqrt{-1}\,\omega^{A*}, \quad \bar\theta^A = \omega^A - \sqrt{-1}\,\omega^{A*},$$

$A = 1, \ldots, n$. Then restricting ω^A's and $\bar\theta^A$'s to M we have

$$\left.\begin{aligned} \theta^\alpha &= \bar\theta^\alpha = \omega^\alpha \quad \text{for} \quad \alpha = p+1, \ldots, p+q, \\ \theta^r &= \bar\theta^r = 0 \quad \text{for} \quad r = p+q+1, \ldots, n. \end{aligned}\right\} \tag{7.11}$$

The Kaehler form $\tilde\Omega$ of N is a closed 2-form on N given by

$$\tilde\Omega = \frac{\sqrt{-1}}{2} \sum_{A=1}^{n} \theta^A \wedge \bar\theta^A. \tag{7.12}$$

Now we can state

THEOREM 7.2 (Chen [7]). Let M be a closed CR-submanifold of a Kaehlerian manifold N. If $H^{2k}(M; R) = 0$ for some $k \leq \dim_C D$, then either D is not integrable or $D^\perp$ is not minimal.

Proof. Let $\Omega = i^*\tilde{\Omega}'$ be the 2-form on M induced from $\tilde{\Omega}$ via the immersion $i : M \to N$. Then from (7.11) and (7.12) we have

$$\Omega = \frac{\sqrt{-1}}{2} \sum_{i=1}^{p} \theta^i \wedge \bar{\theta}^i. \tag{7.13}$$

It is clear that Ω is a closed 2-form on M and it defines a cohomology class $[\Omega]$ in $H^2(M; R)$. Then (7.6) and (7.13) imply that the canonical class $c(M)$ and $[\Omega]$ satisfy

$$[\Omega]^p = (-1)^p (p!) c(M) \tag{7.14}$$

If D is integrable and $D^\perp$ is minimal, then by Theorem 7.1 and (7.14) we obtain $H^{2k}(M; R) \neq 0$ for $k = 1,\ldots,p$. The proof is complete.

Let M be a CR-product of N. Then the distribution D is integrable. Moreover $D^\perp$ is minimal because we have

$$g(X, \nabla_Y Z) = 0 \quad \text{for all} \quad X \in \Gamma(D) \text{ and } Y, Z \in \Gamma(D^\perp).$$

Hence the assumption on the cohomology group in Theorem 7.2 is necessary.

Chapter IV

CR-SUBMANIFOLDS OF COMPLEX SPACE FORMS

§1. Characterization of CR-Submanifolds in Complex Space Forms

Let N(c) be a n-dimensional complex space form of constant holomorphic sectional curvature c. Then the curvature tensor field $\tilde{R}$ of N(C) is given by (see §5 of Chapter I)

$$\tilde{R}(X, Y)Z = \frac{c}{4}\{g(Y, Z)X - g(X, Z)Y + g(JY, Z)JX - g(JX, Z)JY + 2g(X, JY)JZ\}, \tag{1.1}$$

for any vector fields X, Y, Z tangent to N(c).

We give here a characterization of CR-submanifolds in a complex space form in terms of the curvature tensor field of the ambient space.

THEOREM 1.1. (Blair-Chen [1]). *Let M be a submanifold of a complex space form N(c) with c ≠ 0. Then M is a CR-submanifold if and only if the maximal holomorphic subspaces*

$$D_x = T_xM \cap J(T_xM), \quad x \in M$$

define a nontrivial differentiable distribution D on M such that

$$\tilde{R}(D, D, D^{\perp}, D^{\perp}) = \{0\}, \tag{1.2}$$

where $D^{\perp}$ denotes the orthogonal complementary distribution of D in TM.

Proof. Suppose M is a CR-submanifold of N(c). Then by using (1.1) we obtain

$$\tilde{R}(X, Y)Z = \frac{c}{2}\, g(X, JY)JZ,$$

for any X, Y ∈ Γ(D) and Z ∈ Γ($D^{\perp}$). Thus we have (1.2) since JZ is normal to M for any Z ∈ Γ($D^{\perp}$).

Conversely, if the maximal holomorphic subspaces D_x define a nontrivial distribution D such that (1.2) holds, then (1.1) implies

$$0 = \tilde{R}(JX, X, Z, W) = \frac{c}{2}\, g(X, X) g(JZ, W),$$

for any $X \in \Gamma(D)$ and $Z, W \in \Gamma(D^{\perp})$. Thus $JD_x^{\perp}$ is perpendicular to $D_x^{\perp}$. Since D is invariant by J, $JD_x^{\perp}$ is also perpendicular to D_x. Therefore $JD_x^{\perp} \subset T_xM^{\perp}$ and M is a CR-submanifold. The proof is complete.

Taking into account that the curvature tensor field of $N(c)$ is given by (1.1) we have special forms for the structure equations of Gauss, Codazzi and Ricci for the immersion of M in $N(c)$. Thus, the equation of Gauss becomes

$$\begin{aligned} g(R(X, Y)Z, U) &= \frac{c}{4}\{g(Y, Z)g(X, U) - g(X, Z)g(Y, U) + \\ &+ g(\phi Y, Z)g(\phi X, U) - g(\phi X, Z)g(\phi Y, U) + \\ &+ 2g(X, \phi Y)g(\phi Z, W)\} + g(h(Y, Z), h(X, U)) - \\ &- g(h(X, Z), h(Y, U)), \end{aligned} \tag{1.3}$$

for all X, Y, Z tangent to M, where R is the curvature tensor of M. The equation of Codazzi is given by

$$\begin{aligned} (\nabla_X h)(Y, Z) - (\nabla_Y h)(X, Z) &= \frac{c}{4}\{g(\phi Y, Z)\omega X - \\ &- g(\phi X, Z)\omega Y + 2g(X, \phi Y)\omega Z\}, \end{aligned} \tag{1.4}$$

for any X, Y, Z tangent to M. Finally, the Ricci equation becomes

$$\begin{aligned} g(R^{\perp}(X, Y)V, W) &+ g([A_W, A_V]X, Y) = \\ &= \frac{c}{4}\{g(\omega Y, V)g(\omega X, W) - g(\omega X, V)g(\omega Y, W) + \\ &+ 2g(X, \phi Y)g(CV, W)\}, \end{aligned} \tag{1.5}$$

for any X, Y tangent to M and V, W normal to M, where $R^{\perp}$ is the curvature tensor of the normal connection on M.

The special form of the curvature tensor field of $N(c)$ requires a study of the existence of special classes of CR-submanifolds in $N(c)$. Thus we have

THEOREM 1.2 (Bejancu [4]). _There exist no totally umbilical proper CR-submanifolds of $N(c)$ with $c \neq 0$._

Proof. Suppose M is a totally umbilical proper CR-submanifold of N(c) with $c \neq 0$. Let X and Z be two non-null vector fields from D and $D^{\perp}$ respectively. Replacing Y by ϕX in (1.4) and taking into account that M is totally umbilical, the left-hand side of (1.4) becomes

$$g(JX, Z)\nabla^{\perp}_{X}H - g(X, Z)\nabla^{\perp}_{JX}H = 0.$$

On the other hand, the right hand side of (1.4) becomes

$$-\frac{c}{2}\, g(X, X)\omega Z \neq 0.$$

Thus we have a contradiction.

From this theorem we have

COROLLARY 1.1 (Okumura [1]). <u>There exist no totally umbilical real hypersurfaces of N(c) with $c \neq 0$.</u>

COROLLARY 1.2. <u>There exist no totally geodesic proper CR-submanifolds in N(c) with $c \neq 0$.</u>

§2. Riemannian Fibre Bundles and Anti-Holomorphic Submanifolds of CP^n

Let $\tilde{N}$ be a (2n+1)-dimensional regular Sasakian manifold with structure tensors $(\Phi, \xi, \eta, \tilde{g})$. Then there exists a fibering $\tilde{\pi} : \tilde{N} \to \tilde{N}/\xi = N$, where N denotes the set of orbits of ξ and is a real 2n-dimensional Kaehlerian manifold. We denote by (J, g) the Kaehlerian structure on N. Also we denote by * the horizontal lift with respect to the connection η. Then we have

$$(JX)^* = \Phi X^*; \ \tilde{g}(X^*, Y^*) = (g(X, Y))^*, \tag{2.1}$$

for any X, Y tangent to N. We denote by $\tilde{\nabla}'$ (resp. $\tilde{\nabla}$) the Levi-Civita connection with respect to $\tilde{g}$ (resp. g). Then

$$(\tilde{\nabla}_X Y)^* = \tilde{\nabla}'_{X^*}Y^* + \tilde{g}(\Phi X^*, Y^*)\xi = -\Phi^2(\tilde{\nabla}'_{X^*}Y^*). \tag{2.2}$$

We denote by $\tilde{K}$ and $\tilde{R}$ the Riemannian curvature tensors of $\tilde{N}$ and respectively N. Then by using (2.2) we obtain

$$(\tilde{R}(X, Y)Z)^* = \tilde{K}(X^*, Y^*)Z^* + \tilde{g}(Z^*, \Phi Y^*)\Phi X^* - \tilde{g}(Z^*, \Phi X^*)\Phi Y^* - 2\tilde{g}(Y^*, \Phi X^*)\Phi Z^*. \tag{2.3}$$

Now, let $\tilde{M}$ be an (m+1)-dimensional submanifold immersed in $\tilde{N}$ such that the structure vector field ξ of $\tilde{N}$ is tangent to $\tilde{M}$ and let M be an m-dimensional submanifold immersed in N. We assume that there exists a fibration $\pi : \tilde{M} \to M$ such that the diagram

$$\begin{array}{ccc} \tilde{M} & \xrightarrow{i'} & \tilde{N} \\ \pi \downarrow & & \downarrow \tilde{\pi} \\ M & \xrightarrow{i} & N \end{array} \tag{C}$$

is commutative and the immersion i' is a diffeomorphism on the fibres. We denote by the same symbols $\tilde{g}$ and g the induced Riemannian metric on $\tilde{M}$ and M respectively. Let ∇' (resp. ∇) be the Levi-Civita connection on $\tilde{M}$ (resp. M). We denote by $\tilde{h}$ and h the second fundamental forms of the immersions i' and i and by $\tilde{A}$ and A respectively the associated Weingarten maps. Then for any X, Y tangent to M, the Gauss' formulas are given by

$$\tilde{\nabla}_X Y = \nabla_X Y + h(X, Y) \quad \text{and} \quad \tilde{\nabla}'_{X^*} Y^* = \nabla'_{X^*} Y^* + \tilde{h}(X^*, Y^*).$$

From (2.2) we have

$$(\nabla_X Y)^* = -\Phi^2(\nabla'_{X^*} Y^*) \tag{2.4}$$

and

$$(h(X, Y))^* = \tilde{h}(X^*, Y^*). \tag{2.5}$$

Now let $\nabla^\perp$ and $\tilde{\nabla}^\perp$ be the Riemannian connections on the normal bundles to M and $\tilde{M}$ respectively. Then we have Weingarten's formulas

$$\tilde{\nabla}_X V = -A_V X + \nabla^\perp_X V; \quad \tilde{\nabla}'_X V = -\tilde{A}_{V^*} X^* + \tilde{\nabla}^\perp_{X^*} V^*, \tag{2.6}$$

and from (2.2) and (2.6) we obtain

$$(A_V X)^* = -\Phi^2(\tilde{A}_{V^*} X^*) \tag{2.7}$$

and

$$(\nabla^\perp_X V)^* = \tilde{\nabla}^\perp_{X^*} V^*. \tag{2.8}$$

From (6.8) of Chapter I we have

$$(\phi X)^\top = -\nabla'_X \xi; \quad (\Phi X)^\perp = -\tilde{h}(X, \xi), \tag{2.9}$$

for any X tangent to M, where $(\Phi X)^\top$ (resp. $(\Phi X)^\perp$) is the tangent (resp. normal) part of ΦX. From the second equation in (2.9) we obtain $h(\xi, \xi) = 0$.

LEMMA 2.1. <u>Let M be an anti-holomorphic submanifold of a Kaehlerian manifold N and let $\tilde{M}$ be a submanifold of a</u>

Sasakian manifold $\tilde{N}$ satisfying the condition (C). Then we have

$$(\nabla'_{X^*}\tilde{h})(Y^*, Z^*) = ((\nabla_X h)(Y, Z) + g(Y, \phi X)\omega Z + + g(Z, \phi X)\omega Y)^* \tag{2.10}$$

and

$$(\nabla'_{X^*}\tilde{h})(Y^*, \zeta) = (h(Y, \phi X) + h(X, \phi Y))^*, \tag{2.11}$$

for all X, Y, Z tangent to M.

Proof. By using (6.1) of Chapter I, (2.4), (2.5) and (2.7) we obtain

$$((\nabla_X h)(Y, Z))^* = (\nabla'_{X^*}\tilde{h})(Y^*, Z^*) + \tilde{g}(Y^*, \Phi X^*)\cdot \cdot h(\xi, Z^*) + \tilde{g}(Z^*, \Phi X^*)\cdot\tilde{h}(\xi, Y^*).$$

Therefore, by (2.1) and (2.9) we have (2.10). By using (2.9) we obtain

$$(\nabla'_{X^*}\tilde{h})(Y^*, \xi) = -\tilde{\nabla}^{\perp}_{X^*}(\Phi Y^*)^{\perp} + [(\nabla_{X^*}Y^*)]^{\perp} + + \tilde{h}(Y^*, (\Phi X^*)^{T}).$$

Taking into account that $[(\tilde{\nabla}'_{X^*}\Phi)Y^*]^{\perp} = 0$ we obtain

$$\tilde{\nabla}^{\perp}_{X^*}(\Phi Y^*)^{\perp} = (\Phi\nabla'_{X^*}Y^*)^{\perp} - \tilde{h}(X^*, (\Phi Y^*)^{T}).$$

Thus we have

$$(\nabla'_{X^*}\tilde{h})(Y^*, \xi) = \tilde{h}(X^*, (\Phi Y^*)^{T}) + \tilde{h}(Y^*, (\Phi X^*)^{T}),$$

and (2.11) follows by means of (2.1) and (2.5).

THEOREM 2.1 (Yano-Kon [5]). The second fundamental form $\tilde{h}$ of $\tilde{M}$ is parallel if and only if the second fundamental form h of M satisfies the conditions

$$(\nabla_X h)(Y, Z) = g(X, \phi Y)\omega Z + g(X, \phi Z)\omega Y \tag{2.12}$$

and

$$A_V \circ \phi = \phi \circ A_V, \tag{2.13}$$

for any vector fields X, Y, Z tangent to M and any vector field V normal to M.

Proof. By Lemma 2.1 $\tilde{h}$ is parallel on $\tilde{M}$ if and only if (2.12) holds and we have

$$h(X, \phi Y) + h(Y, \phi X) = 0, \tag{2.14}$$

for all $X, Y \in \Gamma(TM)$. The equivalence of (2.13) and (2.14) is a simple verification using (3.3) of Chapter I.

LEMMA 2.2. *Let M be an anti-holomorphic submanifold of a Kaehlerian manifold N and $\tilde{M}$ be a submanifold of a Sasakian manifold $\tilde{N}$ which satisfies the condition (C). Then the normal connection of M is flat if and only if the normal connection of $\tilde{M}$ is flat.*

Proof. By using (2.8) we obtain

$$[g(R^{\perp}(X, Y)V, U)]^* = \tilde{g}(K^{\perp}(X^*, Y^*)V^*, U^*) - $$
$$- 2g(\phi X, Y)^* \cdot g(JV, U)^*, \tag{2.15}$$

for any $X, Y \in \Gamma(TM)$ and $U, V \in \Gamma(TM^{\perp})$, where $R^{\perp}$ and $K^{\perp}$ are the curvature tensor fields of the normal connections on M and $\tilde{M}$ respectively. Since JV is tangent to M, (2.15) becomes

$$[g(R^{\perp}(X, Y)V, U)]^* = \tilde{g}(K^{\perp}(X^*, Y^*)V^*, U^*). \tag{2.16}$$

From (2.9) we obtain $\tilde{A}_{V^*}\xi = (\Phi V^*)^T$. Thus (2.7) implies

$$\tilde{g}([\tilde{A}_{U^*}, \tilde{A}_{V^*}]X, \xi) = (g(h(X, JV), U) -$$
$$- g(h(X, JU), V))^* = (g(A_U JV - A_V JU, X))^*.$$

Taking into account Lemma 2.1 from Chapter III we obtain

$$\tilde{g}([\tilde{A}_{U^*}, \tilde{A}_{V^*}]X^*, \xi) = 0.$$

Hence, by using the Ricci equation for the immersion of $\tilde{M}$ in $\tilde{N}$, we have

$$\tilde{g}(K^{\perp}(X^*, \xi)V^*, U^*) = 0. \tag{2.17}$$

Thus our assertion follows from (2.16) and (2.17).

In the following we assume that M is an m-dimensional submanifold of complex projective space CP^n with constant holomorphic sectional curvature 4 and that $\tilde{M}$ is an $(m+1)$-dimensional submanifold of the unit sphere S^{2n+1} such that the diagram

$$(*) \qquad \begin{array}{ccc} \tilde{M} & \xrightarrow{\ i'\ } & S^{2n+1} \\ \pi \downarrow & & \downarrow \tilde{\pi} \\ M & \xrightarrow{\ i\ } & CP^n \end{array}$$

commutes. Suppose $\dim_C D_x = p$ and $\dim_R D_x^{\perp} = q$. Then we have

LEMMA 2.3. Let M be an m-dimensional anti-holomorphic submanifold of CP^n with flat normal connection and $p \geq 2$. If (2.12) holds then the second fundamental form $\tilde{h}$ of $\tilde{M}$ is parallel.

LEMMA 2.4. Let M be an m-dimensional anti-holomorphic submanifold of CP^n. Then we have

$$g(\nabla h, \nabla h) \geq 4pq, \tag{2.18}$$

and the equality holds if and only if (2.12) holds.

For the proofs of these lemmas see Yano-Kon [3]. In the same paper the authors proved that the diagram

$$\begin{array}{ccc} S^{m_1}(r_1)\times\ldots\times S^{m_k}(r_k) & \xrightarrow{i'} & S^{m+k} \\ \pi \downarrow & & \downarrow \tilde{\pi} \\ \pi(S^{m_1}(r_1)\times\ldots\times S^{m_k}(r_k)) & \xrightarrow{i} & CP^{(m+k-1)/2}, \end{array}$$

$$m + 1 = \sum_{i=1}^{k} m_i, \quad \sum_{i=1}^{k} (r_i)^2 = 1,$$

commutes. Moreover

$$\pi(S^{m_1}(r_1)\times\ldots\times S^{m_k}(r_k))$$

is an example of anti-holomorphic submanifold with flat normal normal connection in a complex projective space and

$$S^{m_1}(r_1)\times\ldots\times S^{m_k}(r_k)$$

is a semi-invariant submanifold (or contact CR-submanifold) in S^{m+k}. We shall give some results on semi-invariant submanifolds in Chapter V.

Now we recall

THEOREM 2.2 (Yano-Kon [3]). Let M be a complete m-dimensional submanifold of S^n with flat normal connection. If the second fundamental form of M is parallel, then M is a small sphere, a great sphere or a pythagorean product of a certain number of spheres. Moreover, if M is of essential codimension n - m then M is a pythagorean product of the form

$$S^{m_1}(r_1)\times\ldots\times S^{m_k}(r_k),\quad \sum_{i=1}^{k}(r_i)^2 = 1,\ k = n-m+1,$$

or a pythagorean product of the form

$$S^{m_1}(r_1)\times\ldots\times S^{m_{k'}}(r_{k'}) \subset S^{n-1} \subset S^n,$$

$$\sum_{i=1}^{k'}(r_i)^2 = r^2 < 1,\ k' = n - m.$$

Then we can state

THEOREM 2.3 (Yano-Kon [3]). Let M be a complete m-dimensional anti-holomorphic submanifold of CP^n with flat normal connection. If $p \geq 2$ and $g(\nabla h, \nabla h) = 4pq$, then M is

$$\pi(S^{m_1}(r_1)\times\ldots\times S^{m_k}(r_k)),\quad m + 1 = \sum_{i=1}^{k} m_i,$$

$$\sum_{i=1}^{k}(r_i)^2 = 1,$$

$2 \leq k \leq m + 1$, where $m_1,\ldots,m_k$ are odd numbers and $n = (m + k - 1)/2$.

Proof. We consider the diagram (*). Since the normal connection of M is flat, by Lemma 2.2 the normal connection of $\tilde{M}$ is flat. Therefore Lemmas 2.3 and 2.4 imply that the second fundamental form of $\tilde{M}$ is parallel. On the other hand, if M admits a geodesic section, that is, if there exists a normal vector V such that $g(h(X, Y), V) = 0$ for all X, Y tangent to M, then (1.5) implies that the holomorphic sectional curvature of the ambient manifold is zero. This is a contradiction. Consequently, the immersion i is full. Thus by (2.5) the immersion i' is full too. Therefore Theorem 2.2 and the example above prove our assertion.

More results on the geometry of anti-holomorphic submanifolds in CP^n via Riemannian fibre bundles are obtained by Yano and Kon in [3]. Moreover, they obtained in [5] and [6] results on CR-submanifolds of CP^n by using the same method of Riemannian fibre bundles.

§3. CR-Products of Complex Space Forms

In §5 of Chapter III we obtained the main results on CR-products of Kaehlerian manifolds. The aim of this paragraph is to study the existence of CR-products in complex space forms.

First, from Theorems 5.4 and 5.5 of Chapter III we have

PROPOSITION 3.1. (i) *There exist no proper CR-products in any complex hyperbolic space N(c), c $<$ 0.*
(ii) *There exist no anti-holomorpnic CR-products in any eliptic complex space form N(c), c $>$ 0.*

The first statement is due to Chen [5] and the second to Bejancu-Kon-Yano [1]. The Proposition 3.1 leads us to a study of CR-products in C^n and to a study of proper CR-products (but not anti-holomorphic) in CP^n.

For CR-products in C^n we have

THEOREM 3.1 (Chen [5]). *Every CR-product in C^n is the Riemannian product of a holomorphic submanifold in a linear complex subspace C^s and a totally real submanifold of C^{n-s} locally, i.e.,*

$$M = M_1 \times M_2 \subset C^s \times C^{n-s} = C^n.$$

Proof. From Lemma 5.1 of Chapter III we obtain

$$h(X, Y) = 0, \tag{3.1}$$

for any $X \in \Gamma(D)$ and $Y \in \Gamma(D^\perp)$. Thus by applying a lemma of Moore [1] we see that $M = M_1 \times M_2$ is a product submanifold in $R^r \times R^{2n-r}$. Since M_1 is a holomorphic submanifold of C^n we may choose R^r to be a complex linear subspace of C^n.

THEOREM 3.2. (Bejancu-Kon-Yano [1]). *Every m-dimensional anti-holomorphic product M of C^n is the Riemannian product $C^p \times M_2$ of C^n, where M_2 is an (2n - m)-dimensional totally real submanifold of C^n.*

Proof. Since M is an anti-holomorphic product, by Corollary 5.1 of Chapter III we have $h(U, V) = 0$ for any $U, V \in \Gamma(D)$. Then by Theorem 1.2 of Chapter III each leaf of D is a totally geodesic submanifold of C^n. Hence each leaf of D is itself a C^p in C^n and our assertion follows.

Now we give some examples of CR-products in CP^n. In order to do this we define a mapping

by
$$S_{pq} : CP^p \times CP^q \to CP^{p+q+pq}$$
$$(z_0, \ldots, z_p; u_0, \ldots, u_q) \to (z_0u_0, \ldots, z_iu_j, \ldots, z_pu_q),$$

where $(z_0, \ldots, z_p)$ (resp. $(u_0, \ldots, u_q)$) are the homogeneous coordinates of CP^p (resp. CP^q). The mapping S is a Kaehler imbedding of the Riemannian product $CP^p \times CP^q$ into C^{p+q+pq}. Let M_2 be a real q-dimensional totally real submanifold of CP^q. Then $CP^p \times M_2$ is a CR-product $M_1 \times M_2$ in CP^{p+q+pq} via S_{pq}, in which $M_1 = CP^p$ is a totally geodesic holomorphic submanifold and M_2 is a totally real submanifold in CP^{p+q+pq}. With these examples in mind we give the following definition.

A CR-product $M_1 \times M_2$ in CP^n is called a <u>standard CR-product</u> if we have:

(i) $n = p + q + pq$, where $p = \dim_C D_x$, $q = \dim_R D_x^\perp$; and

(ii) M_1 is a totally geodesic holomorphic submanifold of CP^n.

We shall prove that in fact $n = p + q + pq$ is the smallest dimension of CP^n for admitting a $(2p+q)$-dimensional CR-product. First we have

LEMMA 3.1. <u>Let M be a CR-product in CP^n. Then</u>

$$\{h(X_i, Z_\alpha)\}, \quad i = 1, \ldots, 2p; \ \alpha = 1, \ldots, q,$$

<u>are orthonormal vectors in</u> ν $(TM^\perp = JD^\perp \oplus \nu)$, <u>where</u> $\{X_1, \ldots, X_{2p}\}$ <u>and</u> $\{Z_1, \ldots, Z_q\}$ <u>are orthonormal basis in</u> D_x <u>and</u> $D_x^\perp$ <u>respectively</u>.

<u>Proof</u>. By Lemma 5.1 of Chapter III we obtain
$$\|h(X, Z)\| = 1, \tag{3.2}$$
for any unit vector fields $X \in \Gamma(D)$ and $Z \in \Gamma(D^\perp)$. Hence we obtain by linearity
$$g(h(X_i, Z), h(X_j, Z)) = 0, \quad i \neq j. \tag{3.3}$$

By Theorem 5.2 of Chapter III we have $h(X, Z) \in \Gamma(\nu)$. Hence if $q = 1$ the lemma is proved.

Now, suppose $q > 1$. Then from (3.3) it follows that

$$g(h(X_i, Z_\alpha), h(X_j, Z_\beta)) + g(h(X_i, Z_\beta), h(X_j, Z_\alpha)) = 0 \tag{3.4}$$

for $i \neq j$ and $\alpha \neq \beta$. Since $M = M_1 \times M_2$ is a CR-product we have

$$R(X_i, X_j, Z_\alpha, Z_\beta) = 0. \tag{3.5}$$

On the other hand, by (1.1) we have

$$\tilde{R}(X_i, X_j, Z_\alpha, Z_\beta) = 0. \tag{3.6}$$

Thus by (1.3), (3.5) and (3.6) we obtain

$$g(h(X_i, Z_\alpha), h(X_j, Z_\beta)) = g(h(X_j, Z_\alpha), h(X_i, Z_\beta)). \tag{3.7}$$

Finally our assertion follows from (3.2), (3.4), and (3.7).

This lemma implies

THEOREM 3.3 (Chen [5]). Let M be a CR-product in CP^n. Then

$$n \geqslant p + q + pq. \tag{3.8}$$

Remark 3.1. Since the standard CR-products satisfy the equality sign in (3.8) the estimate of n in (3.8) is the best possible.

With respect to this topic we give here other two important results due to Chen.

THEOREM 3.4 (Chen [5]). Every $(2p+q)$-dimensional CR-product in CP^n with $n = p + q + pq$ is a standard CR-product.

THEOREM 3.5 (Chen [5]). Let $M = M_1 \times M_2$ be a $(2p+q)$-dimensional CR-product in CP^n. Then

$$\| h \|^2 \geqslant 4pq. \tag{3.9}$$

Now, we can ask whether there exist proper CR-submanifolds in CP^n which are neither real hypersurfaces nor CR-products. We shall give here a procedure due to Shimizu in order to get such CR-submanifolds.

Let G/K be an irreducible Hermitian symmetric space of compact type. Denote by $\pi : S^{2n+1} \to CP^n$ the Hopf fibration.

For a point $p \in S^{2n+1}$ denote by N the K-orbit of p and put $M = \pi(N)$. We say that p is regular if N has the maximal dimension.

THEOREM 3.6 (Shimizu [1]). Let r be the rank of G/K. Suppose p is a regular element of S^{2n+1} and r is greater than one. Then:

(i) M is a proper CR-submanifold in CP^n of codimension $r - 1$;

(ii) M is not a CR-product;

(iii) M has parallel mean curvature;

(iv) M has flat normal connection.

The proof requires some knowledge of symmetric spaces, so we omit it here.

§4. Mixed Foliate CR-Submanifolds of Complex Space Forms

Let M be a CR-submanifold of a Kaehlerian manifold N. Then M is called mixed foliate if D is integrable and M is mixed geodesic, i.e., the second fundamental form h of M satisfies

$$h(X, Y) = 0, \tag{4.1}$$

for any $X \in \Gamma(D)$ and $Y \in \Gamma(D^{\perp})$. The geometry of mixed foliate CR-submanifolds has been studied by Chen in [5] and [6], Bejancu-Kon-Yano in [1] and Bejancu in [4]. In this section we gather some results from the papers quoted above.

First we have

LEMMA 4.1. Let M be a mixed foliate CR-submanifold of a Kaehlerian manifold N. Then

$$A_V \circ \phi + \phi \circ A_V = 0, \tag{4.2}$$

for any vector field V normal to M.

Proof. It is easy to check that (4.2) is equivalent to

$$g(h(X, \phi Y), V) = g(h(Y, \phi X), V), \tag{4.3}$$

for any $X, Y \in \Gamma(TM)$ and $V \in \Gamma(TM^{\perp})$. Since D is integrable we obtain (4.3) for any $X, Y \in \Gamma(D)$ (see Theorem 1.1 of Chapter III). On the other hand, ϕX and ϕY belong to D for any $X, Y \in \Gamma(TM)$. Hence if we take $X \in \Gamma(D^{\perp})$, (4.3) follows from (4.1). In a similar way we have (4.3) for any $Y \in \Gamma(D^{\perp})$.

Concerning the existence of mixed foliate CR-submanifolds in complex space forms we have

THEOREM 4.1 (Bejancu-Kon-Yano [1]). If M is a mixed foliate proper CR-submanifold of a complex space form N(c) then we have $c \leqslant 0$.

Proof. Let $X, Y \in \Gamma(D)$ and $Z \in \Gamma(D^{\perp})$. Then we have

$$(\nabla_X h)(Y, Z) - (\nabla_Y h)(X, Z) = h(X, \nabla_Y Z) - h(Y, \nabla_X Z). \tag{4.4}$$

Take a vector field $V \in \Gamma(JD^{\perp})$ such that $Z = JV = BV$ and obtain

$$\nabla_Y Z = -\phi(A_V Y) + B(\nabla^{\perp}_Y V). \tag{4.5}$$

Thus by (4.3)-(4.5) we have

$$(\nabla_X h)(Y, Z) - (\nabla_Y h)(X, Z) = h(\phi Y, A_V X) + h(X, A_V \phi Y) \tag{4.6}$$

Now let $X = \phi Y$ and using (4.6) and (1.4) we obtain

$$2h(\phi Y, A_V(\phi Y)) = -\frac{1}{2} cg(\phi Y, \phi Y)V.$$

Therefore

$$0 \leqslant 2g(A_V(\phi Y), A_V(\phi Y)) = -\frac{1}{2} cg(\phi Y, \phi Y)g(V, V) \tag{4.7}$$

which proves our assertion.

COROLLARY 4.1. Let M be a mixed foliate CR-submanifold of a complex space form N(c). If $c > 0$ then M is either a holomorphic submanifold or a totally real submanifold of N(c).

By Theorem 4.1 we need only to study the geometry of mixed foliate CR-submanifolds in N(c) with $c \leqslant 0$.

THEOREM 4.2 (Chen [5]). Let M be a CR-submanifold of C^n. Then M is mixed foliate if and only if M is a CR-product.

Proof. Suppose M is a mixed foliate CR-submanifold of C^n. Then we have $c = 0$ in (4.7) and obtain

$$A_V X = 0, \quad \text{for any} \quad V \in \Gamma(JD^{\perp}) \quad \text{and} \quad X \in \Gamma(D).$$

Thus we have

$$Bh(X, Y) = 0, \tag{4.8}$$

for any $X \in \Gamma(D)$ and $Y \in \Gamma(TM)$, Hence, by Theorem 5.2 of

Chapter III, M is a CR-product.

Conversely, suppose M is a CR-product of C^n. Then D is an integrable distribution and (4.1) follows from (4.8), that is, M is mixed geodesic. The proof is complete.

Remark 4.1. Theorem 4.2 has also been proved by Bejancu-Kon-Yano in [1] for anti-holomorphic submanifolds.

In the remaining part of this paragraph we shall study mixed foliate CR-submanifolds in complex hyperbolic spaces. For simplicity, we consider an n-dimensional complex hyperbolic space H with constant holomorphic sectional curvature -4.

First we prove some lemmas due to Chen [6].

LEMMA 4.2. Let M be a mixed foliate CR-submanifold in H. Then for any unit vectors $X \in \Gamma(D)$ and $Z \in \Gamma(D^{\perp})$ we have

$$\|A_{JZ}X\|^2 = 1 \tag{4.9}$$

and

$$\|h\|^2 \geqslant 2pq. \tag{4.10}$$

The equality sign in (4.10) holds if and only if we have:
(a) the leaves of $D^{\perp}$ are totally geodesic in H; and
(b) $\operatorname{Im} h = JD^{\perp}$.

Proof. We put c = -4 in (4.7) and obtain (4.9). Then taking account of (4.9) we obtain (4.10). Again using (4.9) we see that the equality sign in (4.10) holds if and only if we have

$$\operatorname{Im} h = JD^{\perp} \tag{4.11}$$

and

$$A_{JD^{\perp}}D^{\perp} = \{0\}. \tag{4.12}$$

Thus our assertion follows from Theorem 1.5 of Chapter III.

Let M be a mixed foliate CR-submanifold in H and $\bar{M}$ be a leaf of the holomorphic distribution D. Then $\bar{M}$ is a Kaehlerian submanifold of H. We denote by $\bar{h}$, $\bar{\nabla}^{\perp}$,..., etc., the second fundamental form, the normal connection,..., etc., for $\bar{M}$ in H and by h', $\nabla'^{\perp}$,..., etc., the corresponding quantities for $\bar{M}$ in M. Then

$$\bar{h}(X, Y) = h'(X, Y) + h(X, Y), \tag{4.13}$$

for any $X, Y \in \Gamma(T\bar{M})$.

LEMMA 4.3. Let M be a mixed foliate CR-submanifold in H. Then

$$\bar{A}_Z S = A_{JZ}JX, \quad \bar{A}_{JZ}X = A_{JZ}X, \tag{4.14}$$

$$\bar{\nabla}^{\perp}_X JZ = \nabla^{\perp}_X JZ = \omega(\nabla_X Z), \tag{4.15}$$

$$\bar{\nabla}^{\perp}_X Z = \nabla^{\perp\prime}_X Z, \quad A'_Z X = \bar{A}_Z X, \tag{4.16}$$

<u>for any unit vector fields $X \in \Gamma(T\bar{M})$ and $Z \in \Gamma(D^{\perp}{}_{|\bar{M}})$.</u>

<u>Proof</u>. Since $\bar{M}$ is a Kaehlerian submanifold of H, by using (4.13) we obtain

$$g(\bar{A}_Z X, Y) = g(J\bar{h}(X, Y), JZ) = g(h(JX, Y), JZ) = \\ = g(A_{JZ}JX, Y)$$

and

$$g(\bar{A}_{JZ}X, Y) = g(h(X, Y), JZ) = g(A_{JZ}X, Y), \\ , \forall Y \in \Gamma(T\bar{M}).$$

On the other hand, by Proposition 2.4 of Chapter II, the distribution D is invariant with respect to the action of A_{JZ}. Hence we have (4.14).

Next, for any $X \in \Gamma(T\bar{M})$ and $Z \in \Gamma(D^{\perp})$ we have

$$J(\nabla_X Z) = \tilde{\nabla}_X JZ = -A_{JZ}X + \nabla^{\perp}_X JZ. \tag{4.17}$$

Thus we obtain

$$\nabla^{\perp}_X JZ = \omega(\nabla_X Z).$$

Also

$$\tilde{\nabla}_X JZ = - \bar{A}_{JZ}X + \bar{\nabla}^{\perp}_X JZ,$$

which together with (4.17) implies

$$\bar{\nabla}^{\perp}_X JZ = \nabla^{\perp}_X JZ.$$

Thus we have (4.15). Finally, by using equations of Gauss and Weingarten for the immersions of $\bar{M}$ in H, $\bar{M}$ in M and M in H we obtain

$$- \bar{A}_Z X + \bar{\nabla}^{\perp}_X Z = \tilde{\bar{\nabla}}_X Z = \nabla_X Z = - A'_Z X + \nabla^{\perp\prime}_X Z.$$

Hence we have (4.16). The proof is complete.

LEMMA 4.4. <u>Let M be a mixed foliate CR-submanifold of H. Then</u>

$$\operatorname{Im} \bar{h} = D^{\perp} \oplus JD^{\perp} \tag{4.18}$$

and

$$\bar{A}_Z, \bar{A}_{JZ} \in O(2p), \tag{4.19}$$

<u>for any $Z \in \Gamma(D^{\perp})$, where $O(2p)$ is the orthogonal group.</u>

Proof. Let U be a normal vector field in ν. Then by using (1.5) and (4.15) we obtain

$$[\bar{A}_{JZ}, \bar{A}_U] = 0. \tag{4.20}$$

Taking into account that $\bar{M}$ is a Kaehlerian submanifold of H we have

$$\bar{A}_{JU} = J\bar{A}_U = -\bar{A}_U J \quad \text{(see Ogiue [1])}.$$

Thus, by using (4.20), we obtain

$$0 = \bar{A}_{JZ}\bar{A}_U - \bar{A}_U\bar{A}_{JZ} = J\circ(\bar{A}_U\bar{A}_{JZ} + \bar{A}_{JZ}\bar{A}_U).$$

Since J is nonsingular this gives

$$\bar{A}_U\bar{A}_{JZ} + \bar{A}_{JZ}\bar{A}_U = 0. \tag{4.21}$$

Thus from (4.20) and (4.21) we get

$$\bar{A}_{JZ}\bar{A}_U = 0. \tag{4.22}$$

By using (4.9) and (4.14) we obtain

$$\|\bar{A}_Z X\| = \|\bar{A}_{JZ}X\| = 1,$$

for any unit vectors $X \in \Gamma(T\bar{M})$ and $Z \in \Gamma(D^\perp)$. Thus we have (4.19). In particular $\bar{A}_{JZ}$ is nonsingular. Hence from (4.22) we have $\bar{A}_U = 0$ which implies (4.18).

Now we can state

THEOREM 4.3. *Let M be a mixed foliate CR-submanifold of H. Then M is an anti-holomorphic submanifold of a complex (p+q)-dimensional complex submanifold H' which is totally geodesic in H.*

Proof. Taking into account that $\bar{A}_U = 0$ for each $U \in \Gamma(\nu)$ we obtain

$$\|h\|^2 = \sum_{\alpha,i}\{\|\bar{A}_{Z_\alpha}E_i\|^2 + \|\bar{A}_{JZ_\alpha}E_i\|^2\},$$

where $\{E_1,\ldots,E_{2p}\}$ (resp. $\{Z_1,\ldots,Z_q\}$) is a basis of $T_x\bar{M}$ (resp. $D_x^\perp$). Then by using (4.9) and (4.14) we have

$$\|\bar{h}\|^2 = \sum_{\alpha,i}\{\|A_{JZ_\alpha}JE_i\|^2 + \|A_{JZ_\alpha}E_i\|^2\} = 4pq. \tag{4.23}$$

Next, from (4.9), (4.14), and (4.16) it follows that

$$\|h'\|^2 = \sum_{\alpha,i}\{\|A'_{Z_\alpha}E_i\|^2\} = \sum_{\alpha,i}\{\|A_{JZ_\alpha}JE_i\|^2\} = 2pq. \tag{4.24}$$

Thus, from (4.13) we obtain

$$\|h\|^2 = 2pq. \tag{4.25}$$

Then by using Lemma 4.2 and the equation of Gauss and Weingarten for the immersion of M in H we get

$$J(\nabla_Y Z) = \tilde{\nabla}_Y JZ = -A_{JZ}Y + \nabla^{\perp}_Y JZ,$$

for any $Y \in \Gamma(TM)$ and $Z \in \Gamma(D^{\perp})$. Hence

$$\nabla^{\perp}_Y JZ = \omega(\nabla_Y Z). \tag{4.26}$$

By (4.26) we see that $JD^{\perp}$ is a parallel subbundle of the normal bundle of M in H. Thus, by assertion (b) of Lemma 4.2, we see that M lies in a complex (p+q)-dimensional complex submanifold H' which is totally geodesic in H. Obviously, M is an anti-holomorphic submanifold of H'.

LEMMA 4.5. Let M be a mixed foliate proper CR-submanifold of a complex space form $M(c)$, $c \neq 0$. Then $c < 0$ and $q > 1$.

Proof. By Theorem 4.1 we have $c < 0$. If $q = 1$ then for any unit vector fields $Z \in \Gamma(D^{\perp})$ and $X \in \Gamma(TM)$ by (4.16) we obtain

$$\bar{\nabla}^{\perp}_X Z = \bar{\nabla}^{\perp\prime}_X Z = 0.$$

Hence Z is a parallel normal vector field of the complex submanifold $\bar{M}$ in $N(c)$, $c < 0$. This contradicts a theorem of Chen-Ogiue [2].

PROPOSITION 4.1. Let M be a mixed foliate proper CR-submanifold of H. Then each leaf $\bar{M}$ of D is an Einstein-Kaehler submanifold of H with Ricci tensor $\bar{S}$ given by

$$\bar{S}(X, Y) = -2(p+q+1)g(X, Y). \tag{4.27}$$

Moreover, we have $p > 1$ and $q > 1$.

Proof. Since $\bar{M}$ is a Kaehlerian submanifold of H, by a direct computation we obtain

$$\bar{S}(X, Y) = -2(q+1)g(X, Y) - \sum_{\alpha}\{g(\bar{A}_{\zeta_\alpha}X, \bar{A}_{\zeta_\alpha}Y)\},$$

for any $X, Y \in \Gamma(T\bar{M})$, where $\{\zeta_\alpha\}$ is an orthonormal basis of

$T\bar{M}^{\perp}$. Thus, taking into account that $\bar{A}_U = 0$ for any $U \in \Gamma(\nu)$ and using (4.9) and (4.14) we obtain (4.27). By Lemma 4.5 we have $q > 1$. It remains to prove $p > 1$. Suppose $p = 1$. Then by (4.27) $\bar{M}$ is of constant curvature $-2(q+2)$. Hence a theorem of Calabi [1] gives $q = 0$ which is a contradiction.

By using Theorem 4.3 and taking into account that for a proper mixed foliate CR-submanifold of H we have $p > 1$ and $q > 1$ we obtain

THEOREM 4.4. Let M be a mixed foliate CR-submanifold of H. If $\dim_R M \leqslant 5$ then M is a totally real submanifold.

Thus we have

COROLLARY 4.2. There exist no mixed foliate proper CR-submanifolds of dimension $\leqslant 5$ in a hyperbolic complex space form.

Remark 4.2. It is an open problem if Corollary 4.2 is valid for any dimension of M. If M is a compact manifold, Chen obtained in [10] theorems on non-integrability of the holomorphic distribution.

§5. CR-Submanifolds with Semi-Flat Normal Connection

Let M be a CR-submanifold of a complex space form N(c). Then we say that M has semi-flat normal connection if its normal curvature $R^{\perp}$ satisfies

$$R^{\perp}(X, Y, V, W) = \frac{c}{2} g(X, \phi Y) g(JV, W), \tag{5.1}$$

for any $X, Y \in \Gamma(TM)$ and $V, W \in \Gamma(TM^{\perp})$. By using (1.5) and (5.1) we obtain that a CR-submanifold M of a complex space form N(c) has semi-flat normal connection if and only if

$$g([A_W, A_V]X, Y) = \frac{c}{4}\{g(\omega Y, V) g(\omega X, W) - g(\omega X, V) g(\omega Y, W)\}. \tag{5.2}$$

CR-submanifolds with semi-flat normal connection have been defined by Yano and Kon in [5]. A complete classification of CR-submanifolds with semi-flat normal connection immersed in N(c) with $c \neq 0$ has been obtained by Chen in [6]. More precisely we have

THEOREM 5.1. Let M be a CR-submanifold in a complex space form N(c), $c \neq 0$. Then M has semi-flat normal connection in N(c) if and only if M is one of the following:

(i) a totally geodesic complex submanifold;

(ii) a flat totally real submanifold of a totally geodesic complex submanifold $\tilde{N}(c)$ in N(c);

(iii) a proper anti-holomorphic submanifold with flat normal connection in a totally geodesic complex submanifold N'(c) of N(c);

(iv) a space of positive constant sectional curvature immersed in a totally geodesic complex submanifold N"(c) of N(c) with flat normal connection as a totally real submanifold.

CR-submanifolds with semi-flat normal connection in CP^n have been studied by Yano and Kon in [5] and [6]. The method of Riemannian fibre bundles was also the main tool in obtaining interesting results. We only mention here a theorem which is a generalization of Theorem 2.3.

THEOREM 5.2 (Yano-Kon [5]). Let M be a complete m-dimensional CR-submanifold of CP^n with semi-flat normal connection and $p > 1$. If the f-structure C in the normal bundle to M is parallel and if $g(\nabla h, \nabla h) = 4pq$, then M is a totally geodesic holomorphic submanifold $CP^{m/2}$ of CP^n or M is an anti-holomorphic submanifold of $CP^{(m+q)/2}$ in CP^n and it is

$$\pi(S^{m_1}(r_1)\times\ldots\times S^{m_k}(r_k)), \quad m + 1 = \sum_{i=1}^{k} m_i,$$
$$\sum_{i=1}^{k} (r_i)^2 = 1,$$

where $m_1,\ldots,m_k$ are odd numbers.

More results on the geometry of CR-submanifolds with semi-flat normal connection can be found in Yano-Kon [6] and Chen [6].

§6. Pinching Theorems for Sectional Curvatures of CR-Submanifolds

Let M be a real m-dimensional CR-submanifold in an n-dimensional complex space form N(c). A plane section Δ of the tangent bundle TM is called an anti-holomorphic plane section if $J(\Delta)$ is orthogonal to Δ. Then the sectional curvature $K(\Delta)$ of M is called an anti-holomorphic sectional curvature of M. If the plane section belongs to the holomorphic distribution we say that $K(\Delta)$ is a D-anti-holomorphic sectional curvature of M. The distribution $D^{\perp}$ is called D-parallel if we have

$$\nabla_X Y \in \Gamma(D^{\perp}), \tag{6.1}$$

for all $X \in \Gamma(D)$ and $Y \in \Gamma(D^{\perp})$.

LEMMA 6.1. Let M be a CR-submanifold of N(c). Then the distribution $D^{\perp}$ is D-parallel if and only if

$$g(h(X, Y), JZ) = 0, \tag{6.2}$$

for any $X, Y \in \Gamma(D)$ and $Z \in \Gamma(D^{\perp})$.

Proof. Since g is Hermitian, by (3.2) of Chapter III we obtain

$$g(h(X, Y), JZ) = -g(Bh(X, Y), Z) = -g((\nabla_X \phi)Y, Z).$$

Since ∇ is a Riemannian connection we get

$$g(h(X, Y), JZ) = g(\phi Y, \nabla_X Z),$$

and the lemma follows.

LEMMA 6.2. Let M be a CR-submanifold of N(c). If $D^{\perp}$ is D-parallel then the holomorphic distribution is integrable and its leaves are totally geodesic in M.

The proof of the lemma follows from Theorem 1.2 of Chapter III and Lemma 6.1.

Now let $\{V_1, \ldots, V_{2n-m}\}$ be a local field of orthonormal frames in the normal bundle $TM^{\perp}$ and let $A_\alpha = A_{V_\alpha}$ $(\alpha = 1, \ldots, 2n - m)$ be the fundamental tensors of Weingarten. The mean curvature vector H is given by

$$H = \frac{1}{m} \sum_\alpha \mathrm{Tr}(A_\alpha) V_\alpha.$$

Then by a direct computation the scalar curvature ρ of M is given by

$$\rho = \frac{1}{4}(m^2 - m + 6p)c + m^2 \|H\|^2 - \|h\|^2, \tag{6.3}$$

where we have

$$\|H\|^2 = \frac{1}{m^2} \sum_\alpha (\mathrm{Tr}(A_\alpha))^2 \quad \text{and} \quad \|h\|^2 = \sum_\alpha \mathrm{Tr}(A_\alpha^2)$$

Also, the sectional curvature of M determined by orthonormal vectors $\{X, Y\}$ is given by

$$K_M(X \wedge Y) = \frac{c}{4}\{1 + 3g(PX, \phi Y)^2\} + g(h(X, X), h(Y, Y)) - \\ - g(h(X, Y), h(X, Y)). \tag{6.4}$$

Let $\{E_1,\ldots,E_m\}$ be a local field of orthonormal frames tangent to M. Then we put

$$h(E_i, E_j) = \sum_{\alpha} h^{\alpha}_{ij} V_{\alpha}, \quad i, j = 1,\ldots,m. \tag{6.5}$$

By using (6.4) and (6.5) we obtain

$$\|h\|^2 = \sum_{\alpha,i,j} (h^{\alpha}_{ij})^2 \tag{6.6}$$

and

$$K_M(E_i \wedge E_j) = \frac{c}{4}\{1 + 3g(PE_i, \phi E_j)^2\} + \\ + \sum_{\alpha} \{h^{\alpha}_{ij} h^{\alpha}_{jj} - (h^{\alpha}_{ij})^2\}, \tag{6.7}$$

for all $i \neq j$.

Now we can state

THEOREM 6.1 (Bejancu [18]). Let M be a CR-submanifold of a complex space form N(c). Suppose the following conditions are fulfilled:

(i) the distribution $D^{\perp}$ is D-parallel;

(ii) there exists $k > 0$ such that the second fundamental form h of M satisfies

$$\|h\|^2 \leqslant k. \tag{6.8}$$

Then all the D-anti-holomorphic sectional curvatures of M are $\geqslant (c - k)/4$.

Proof. Let Δ be an anti-holomorphic plane section which belongs to the distribution D. Suppose $\{E_1, E_2\}$ is an orthonormal field of frames of Δ. Then choose the orthonormal field of frames

$$\{E_1,\ldots,E_p, E_{1*} = JE_1,\ldots,E_{p*} = JE_p, F_1,\ldots,F_q, \\ F_{1*} = JF_1,\ldots,F_{q*} = JF_q, V_1,\ldots,V_s, \\ V_{1*} = JV_1,\ldots,V_{s*} = JV_s\} \tag{6.9}$$

on N(c) such that $E_i \in \Gamma(D)$ $(i = 1,\ldots,p)$, $F_a \in \Gamma(D^{\perp})$ $(a = 1,\ldots,q)$ and $V_{\alpha} \in \Gamma(\nu)$ $(\alpha = 1,\ldots,s)$. By using this field of frames, from (6.7) we obtain

$$K_M(E_1 \wedge E_2) = K_{12} = \frac{c}{4} + \sum_{\alpha=1}^{s} \{h_{11}^{\alpha} h_{22}^{\alpha} + h_{11}^{\alpha*} h_{22}^{\alpha*} - (h_{12}^{\alpha})^2 - (h_{12}^{\alpha*})^2\} + \sum_{a=1}^{q} \{h_{11}^{a} h_{22}^{a} - (h_{12}^{a})^2\} \quad (6.10)$$

Taking account of Lemmas 6.1 and 6.2 we have

$$h_{ij}^{\alpha} = -h_{i*j*}^{\alpha}; \quad h_{ij}^{\alpha*} = -h_{i*j*}^{\alpha*} \quad (6.11)$$

and

$$h_{ij}^{a*} = 0. \quad (6.12)$$

Hence by (6.8), (6.11) and (6.12) it follows that

$$k \geqslant \|h\|^2 \geqslant 2 \sum_{\alpha=1}^{s} \{(h_{11}^{\alpha})^2 + (h_{22}^{\alpha})^2 + 2(h_{12}^{\alpha})^2 + (h_{11}^{\alpha*})^2 + (h_{22}^{\alpha*})^2 + 2(h_{12}^{\alpha*})^2\} \geqslant$$

$$\geqslant 4 \sum_{\alpha=1}^{s} \{(h_{12}^{\alpha})^2 + (h_{12}^{\alpha*})^2 - h_{11}^{\alpha} h_{22}^{\alpha} - h_{11}^{\alpha*} h_{22}^{\alpha*}\}. \quad (6.13)$$

Finally, from (6.10) and (6.13) and using (6.12) we obtain $K_{12} \geqslant (c - k)/4$, that is, the assertion of the theorem.

Next we need an algebraic lemma

LEMMA 6.3 (Okumura [2]). *Let $a_1, \ldots, a_m$ and t be real numbers satisfying the inequality*

$$(\sum_i a_i)^2 \geqslant (m - 1) \{\sum_i (a_i)^2 + t\}.$$

Then for any pair of distinct i and j we have $2a_i a_j \geqslant t$.

Now we state

THEOREM 6.2 (Bejancu [18]). *Let M be a CR-submanifold of a complex space form $N(c)$, $c \geqslant 0$. If the scalar curvature ρ of M satisfies*

$$\rho \geqslant \frac{1}{4}(m^2 - 3m + 6p + 2)c + (m - 2) \cdot \|h\|^2 + 2(m - 1)k, \quad (6.14)$$

then all sectional curvatures of M are $\geqslant k$.

Proof. We note first that from (6.4) we have

$$K_M(X \wedge Y) \geqslant \frac{c}{4} + g(h(X, X), h(Y, Y)) - $$
$$- g(h(X, Y), h(X, Y)) \tag{6.15}$$

for all orthonormal vectors X, Y tangent to M. Now, by (6.3) and (6.14) we obtain

$$m^2 \cdot \|H\|^2 \geqslant (m - 1)\{\|h\|^2 + 2k - \frac{c}{2}\}; \tag{6.16}$$

Since we have to prove a pointwise theorem, we shall take an arbitrary point $x \in M$. Then we have either $\|H\| \neq 0$ or M is minimal at x.

Suppose $\|H\| \neq 0$ at x. Then we can choose the field of orthonormal frames $\{V_1, \ldots, V_{2n-m}\}$ in $TM^{\perp}$ such that

$$V_1 = \frac{1}{\|H\|} \cdot H.$$

Hence (6.16) becomes

$$\left(\sum_{i=1}^{m} h^1_{ii}\right)^2 \geqslant (m - 1)\{\|h\|^2 + 2k - \frac{c}{2}\}. \tag{6.17}$$

Next, by using Lemma 6.3 and a computation similar to the one made by Chen and Okumura for the proof of Theorem 1 of [1] we obtain

$$\frac{c}{4} + \sum_{\alpha=1}^{2n-m} \{h^\alpha_{ii} h^\alpha_{jj} - (h^\alpha_{ij})^2\} \geqslant k, \tag{6.18}$$

for arbitrary $i \neq j$ $(i, j = 1, \ldots, m)$. Thus, in this case the theorem follows from (6.15) and (6.18).

If $H = 0$ at x, then from (6.16) we have

$$k \leqslant \frac{c}{4} - \frac{1}{2}\|h\|^2 \leqslant \frac{c}{4} - \frac{1}{2} \sum_{\alpha=1}^{2n-m} \{(h^\alpha_{ii})^2 + (h^\alpha_{jj})^2 +$$
$$+ 2(h^\alpha_{ij})^2\} \leqslant \frac{c}{4} + \sum_{\alpha=1}^{2n-m} \{h^\alpha_{ii} h^\alpha_{jj} - (h^\alpha_{ij})^2\}, \tag{6.19}$$

for all $i \neq j$. Hence by (6.19) and (6.15) the proof of the theorem is complete.

Chapter V

EXTENSIONS OF CR-STRUCTURES TO OTHER GEOMETRICAL STRUCTURES

§1. Semi-Invariant Submanifolds of Sasakian Manifolds

Let N be a real $(2n + 1)$-dimensional almost contact metric manifold with structure tensors (ϕ, ξ, η, g), where ϕ is a tensor field of type $(1, 1)$, ξ is a vector field, η is a 1-form and g is a Riemannian metric on N. These tensor fields are related by (see §6 of Chapter I)

$$\phi^2X = -X + \eta(X)\xi;\ \phi\xi = 0;\ \eta(\xi) = 1;\ \eta(\phi X) = 0 \quad (1.1$$

$$g(\phi X, \phi Y) = g(X, Y) - \eta(X)\cdot\eta(Y);\ \eta(X) = g(X, \xi), \quad (1.2)$$

for any vector fields X, Y tangent to N.

Let M be a real m-dimensional submanifold of N. We assume that the vector field ξ is tangent to M. Denote by $\{\xi\}$ the 1-dimensional distribution spanned by ξ on M. Then M is called a <u>semi-invariant submanifold</u> of N if there exist two differentiable distributions D and $D^\perp$ on M satisfying

(i) $TM = D \oplus D^\perp \oplus \{\xi\}$,

where D, $D^\perp$ and $\{\xi\}$ are mutually orthogonal to each other;

(ii) the distribution D is invariant by ϕ, that is, $\phi(D_x) = D_x$ for each $x \in M$;

(iii) the distribution $D^\perp$ is anti-invariant by ϕ, that is, $\phi(D_x^\perp) \subset T_xM^\perp$ for each $x \in M$.

<u>Remark 1.1</u>. The study of semi-invariant submanifolds in Sasakian manifolds has been initiated by Bejancu-Papaghiuc in [1]. The same concept was studied under the name <u>"contact CR-submanifold"</u> by Yano-Kon [7], [8], Kobayashi M. [1]-[4], Hsu [1], Matsumoto [2], Arca-Roşca [1].

<u>Remark 1.2</u>. A semi-invariant submanifold is nothing but the extension of the concept of a CR-submanifold of a Kaehler manifold to submanifolds of almost contact metric manifolds.

We denote by 2p and q the real dimensions of D_x and $D_x^{\perp}$ respectively, $x \in M$. Then we see that for p = 0 we obtain an anti-invariant submanifold tangent to ξ (see Yano-Kon [1]) and for q = 0 we obtain an invariant submanifold (see Kon [2]). On the other hand, it is easy to check that each hypersurface of N which is tangent to ξ inherits the structure of a semi-invariant submanifold of N.

In this paragraph we are concerned with semi-invariant submanifolds of Sasakian manifolds (see §6 of Chapter I for Sasakian manifolds). More precisely, we shall outline a study of the integrability of the distributions on M and of the immersion of their leaves in M or N.

Let M be a semi-invariant submanifold of a Sasakian manifold N. Then

$$(\tilde{\nabla}_X \phi)Y = g(X, Y)\xi - \eta(Y)X \tag{1.3}$$

and

$$\tilde{\nabla}_X \xi = -\phi X, \tag{1.4}$$

for any X, Y tangent to N, where $\tilde{\nabla}$ is the Levi-Civita connection on N. We denote by P and Q the projection morphisms of TM on D and $D^{\perp}$ respectively. Then

$$X = PX + QX + \eta(X)\xi, \tag{1.5}$$

for any $X \in \Gamma(TM)$. Also we put

$$\phi V = BV + CV, \tag{1.6}$$

for any $V \in \Gamma(TM^{\perp})$, where BV is the tangent part of ϕV and CV is the normal part of ϕV. We define two tensor fields $\psi : TM \to TM$ and $\omega : TM \to TM^{\perp}$ by

$$\psi X = \phi PX \tag{1.7}$$

and respectively

$$\omega X = \phi QX, \quad \text{for any } X \in \Gamma(TM). \tag{1.8}$$

Now, by using (1.1)-(1.8) and the equations of Gauss and Weingarten for the immersion of M in N, we obtain the following lemmas (see Bejancu-Papaghiuc [1]).

LEMMA 1.1. <u>Let M be a semi-invariant submanifold of a Sasakian manifold N. Then</u>

$$\nabla_X \psi Y - A_{\omega Y} X = \psi(\nabla_X Y) + Bh(X, Y) - \eta(Y)X, \tag{1.9}$$

$$\eta(\nabla_X \psi Y) = g(\phi X, \phi Y) + \eta(A_{\omega Y} X), \tag{1.10}$$

$$h(X, \psi Y) + \nabla^{\perp}_{X}\omega Y = Ch(X, Y) + \omega(\nabla_X Y), \tag{1.11}$$

for any $X, Y \in \Gamma(TM)$.

LEMMA 1.2. Let M be a semi-invariant submanifold of a Sasakian manifold N. Then

$$h(X, \xi) = 0, \quad \nabla_X \xi = -\phi X, \quad \text{for any} \quad X \in \Gamma(D) \tag{1.12}$$

and

$$h(Y, \xi) = -\phi Y, \quad \nabla_Y \xi = 0, \quad \text{for any} \quad Y \in \Gamma(D^{\perp}). \tag{1.13}$$

LEMMA 1.3. Let M be a semi-invariant submanifold of a Sasakian manifold N. Then

$$A_{\phi X} Y = A_{\phi Y} X \tag{1.14}$$

and

$$[X, Y] \in \Gamma(D \oplus D^{\perp}), \quad \text{for any} \quad X, Y \in \Gamma(D^{\perp}). \tag{1.15}$$

Remark 1.3. In these lemmas we have made use of notations from §3 of Chapter I with respect to the immersion of M in N.

Remark 1.4. From (1.13) it follows that a totally umbilical semi-invariant submanifold of a Sasakian manifold is an invariant submanifold.

By using these lemmas we obtain the covariant derivatives of ψ, ω, B and C.

PROPOSITION 1.1. Let M be a semi-invariant submanifold of a Sasakian manifold N. Then

$$(\nabla_X \psi) Y = Bh(X, Y) + A_{\omega Y} X + g(X, Y)\xi - \eta(Y)X, \tag{1.16}$$

$$(\nabla_X \omega) Y = Ch(X, Y) - h(X, \psi Y), \tag{1.17}$$

$$(\nabla_X B) V = A_{CV} X - \psi(A_V X), \tag{1.18}$$

$$(\nabla_X C) V = -\omega(A_V X) - h(X, BV), \tag{1.19}$$

for any $X, Y \in \Gamma(TM)$ and $V \in \Gamma(TM^{\perp})$.

Now we can state

THEOREM 1.1 (Bejancu-Papaghiuc [1]). Let M be a semi-invariant submanifold of a Sasakian manifold N. Then the distribution $D^{\perp}$ is integrable.

Proof. Take $X, Y \in \Gamma(D^{\perp})$ and using (1.3) and the equations of Gauss and Weingarten we obtain

$$\phi(\nabla_Y X) + \phi h(X, Y) + g(X, Y)\xi =$$

$$= -A_{\phi X}Y + \nabla^{\perp}_Y \phi X. \tag{1.20}$$

By changing X and Y in (1.20) and then subtracting the obtained relation from (1.20) we have

$$\phi([Y, X]) = A_{\phi Y}X - A_{\phi X}Y + \nabla^{\perp}_Y \phi X - \nabla^{\perp}_X \phi Y. \tag{1.21}$$

By (1.1) and Lemma 1.3, (1.21) becomes

$$[X, Y] = B(\nabla^{\perp}_Y \phi X - \nabla^{\perp}_X \phi Y). \tag{1.22}$$

Since $BV \in \Gamma(D^{\perp})$ for any $V \in \Gamma(TM^{\perp})$, the assertion follows from (1.22).

Using (1.3) and Lemma 1.2 we obtain

$$g([X, \xi], Y) = 0, \tag{1.23}$$

for any $X \in \Gamma(D^{\perp})$ and $Y \in \Gamma(D)$. Hence, by Theorem 1.1, we have

COROLLARY 1.1. Let M be a semi-invariant submanifold of a Sasakian manifold N. Then the distribution $D^{\perp} \oplus \{\xi\}$ is integrable.

Now we state

LEMMA 1.4. Let M be a semi-invariant submanifold of a Sasakian manifold N. Then

$$g(h(X, Y), \phi Z) = g(\nabla_X Z, \phi Y) \tag{1.24}$$

and

$$[Y, \xi] \in \Gamma(D \oplus \{\xi\}), \tag{1.25}$$

for any $X \in \Gamma(TM)$, $Y \in \Gamma(D)$ and $Z \in \Gamma(D^{\perp})$.

Proof. Using the equations of Gauss and Weingarten we obtain (1.24). Next, by (1.4) and Lemma 1.2 we have

$$g([Y, \xi], Z) = -g(\tilde{\nabla}_{\xi}Y, Z) = g(\nabla_{\xi}Y, Z) =$$

$$= g(Y, \nabla_{\xi}Z), \tag{1.26}$$

for each $Y \in \Gamma(D)$ and $Z \in \Gamma(D^{\perp})$. Now we take $X \in \Gamma(D)$ such that $Y = \phi X$ and using (1.24) and (1.12) obtain

$$g(Y, \nabla_{\xi}Z) = g(\phi X, \nabla_{\xi}Z) = g(h(\xi, X), \phi Z) = 0. \tag{1.27}$$

Thus, by (1.26) and (1.27), we have (1.25).

THEOREM 1.2 (Bejancu-Papaghiuc [1]). Let M be a semi-invariant submanifold of a Sasakian manifold N. Then the distribution $D \oplus \{\xi\}$ is integrable if and only if

$$g(h(X, \phi Y) - h(Y, \phi X), \phi Z) = 0, \tag{1.28}$$

for all $X, Y \in \Gamma(D)$ and $Z \in \Gamma(D^{\perp})$.

Proof. By (1.11) we have

$$h(X, \phi Y) = Ch(X, Y) + \psi(\nabla_X Y), \tag{1.29}$$

for all $X, Y \in \Gamma(D)$. Since h is symmetric, if follows that

$$h(X, \phi Y) - h(Y, \phi X) = \omega([X, Y]). \tag{1.30}$$

Thus $[X, Y] \in \Gamma(D \oplus \{\xi\})$ if and only if (1.28) is satisfied. Therefore, by (1.25) the proof is complete.

By a direct computation we obtain

$$g([X, Y], \xi) = 2g(Y, \phi X),$$

for any $X, Y \in \Gamma(D)$. Hence we have

PROPOSITION 1.2. Let M be a semi-invariant submanifold of a Sasakian manifold N with $D \neq \{0\}$. Then the distributions D and $D \oplus D^{\perp}$ are not integrable.

From these results we conclude that it might be interesting to study the geometry of the leaves of the distributions $D \oplus \{\xi\}$ and $D^{\perp}$.

First we state

THEOREM 1.3. Let M be a semi-invariant submanifold of a Sasakian manifold N. Then

(i) The distribution $D \oplus \{\xi\}$ is integrable and its leaves are totally geodesic in M if and only if

$$g(h(X, Y), \phi Z) = 0, \tag{1.31}$$

for any $X, Y \in \Gamma(D \oplus \{\xi\})$;

(ii) The distribution $D \oplus \{\xi\}$ is integrable and its leaves are totally geodesic in N if and only if

$$h(X, Y) = 0, \tag{1.32}$$

for any $X, Y \in \Gamma(D \oplus \{\xi\})$.

Proof. The distribution $D \oplus \{\xi\}$ is integrable and its leaves are totally geodesic in M if and only if $\nabla_X Y \in \Gamma(D \oplus \{\xi\})$ for any $X, Y \in \Gamma(D \oplus \{\xi\})$. Thus the assertion (i) follows by using (1.29) and (1.12).

Now suppose $D \oplus \{\xi\}$ is integrable and its leaves are totally geodesic in N. Then we have $\tilde{\nabla}_X Y \in \Gamma(D \oplus \{\xi\})$ for any $X, Y \in \Gamma(D \oplus \{\xi\})$. By using the equation of Gauss we obtain

$$g(h(X, Y), V) = g(\tilde{\nabla}_X Y, V) = 0,$$

for any $V \in \Gamma(TM^\perp)$, that is, (1.32) is satisfied.

Conversely, suppose (1.32) is satisfied. Then by Theorem 1.2, $D \oplus \{\xi\}$ is integrable. Let M* be a leaf of $D \oplus \{\xi\}$ and denote by h* (resp. h') the second fundamental form of the immersion of M* in N (resp. M). Then by using (1.32) we have $h^* = h'$. On the other hand, by the assertion (i) M* is totally geodesic in M. Hence $h^* = h' = 0$, that is, M* is totally geodesic in N.

THEOREM 1.4 (Bejancu-Papaghiuc [1]). Let M be a semi-invariant submanifold of a Sasakian manifold N. Then any leaf of $D^\perp$ is totally geodesic in M if and only if

$$g(h(X, Y), \phi Z) = 0, \tag{1.33}$$

for any $Y \in \Gamma(D)$ and $X, Z \in \Gamma(D^\perp)$.

Proof. By using (1.9) we obtain

$$g(\psi(\nabla_X Z), Y) = -g(h(X, Y), \phi Z). \tag{1.34}$$

Let $\bar{M}$ be a leaf of $D^\perp$. Denote by $\bar{\nabla}$ the Levi-Civita connection on $\bar{M}$ and by $\bar{h}$ the second fundamental form of the immersion of $\bar{M}$ in M. Hence the Gauss formula is given by

$$\nabla_X Z = \bar{\nabla}_X Z + \bar{h}(X, Z), \tag{1.35}$$

for any $X, Z \in \Gamma(T\bar{M})$. By using (1.34) and (1.35) we obtain

$$g(\bar{h}(X, Z), \phi Y) = g(h(X, Y), \phi Z). \tag{1.36}$$

On the other hand, by (1.13) we have

$$\eta(\nabla_X Z) = g(\nabla_X Z, \xi) = -g(Z, \nabla_X \xi) = 0.$$

Thus by (1.35) we obtain

$$\eta(\bar{h}(X, Z)) = 0. \tag{1.37}$$

Therefore, the assertion follows from (1.36) and (1.37).

Now we say that M is a $(D, D^\perp)$-geodesic (resp. $D^\perp$-geodesic) semi-invariant submanifold if

$$h(D, D^{\perp}) = \{0\} \quad (\text{resp. } h(D^{\perp}, D^{\perp}) = \{0\}). \tag{1.38}$$

Then by Theorem 1.4 we obtain

COROLLARY 1.2. Let M be a $(D, D^{\perp})$-geodesic semi-invariant submanifold of a Sasakian manifold N. Then we have

(i) each leaf of $D^{\perp}$ is totally geodesic in M;

(ii) each leaf of $D^{\perp}$ is totally geodesic in N if and only if M is $D^{\perp}$-geodesic.

We denote by ν the complementary subbundle orthogonal to $\phi D^{\perp}$ in $TM^{\perp}$. Then we state

THEOREM 1.5 (Bejancu-Papaghiuc [1]). Let M be a semi-invariant submanifold of a Sasakian manifold N. Then any leaf of $D^{\perp}$ is totally geodesic in M if and only if

(i) $\nabla^{\perp}_X \phi Y \in \Gamma(\phi D^{\perp})$, for all $X, Y \in \Gamma(D^{\perp})$; and

(ii) $h(X, Z) \in \Gamma(\nu)$, for all $X \in \Gamma(D^{\perp})$ and $Z \in \Gamma(D \oplus D^{\perp})$.

Proof. Adding (1.9) and (1.11) we obtain

$$\begin{aligned} h(X, Y) &= \omega(A_{\omega Y}X) - \phi(\nabla^{\perp}_X \omega Y) \\ &= \omega(A_{\omega Y}X) - C(\nabla^{\perp}_X \omega Y), \end{aligned} \tag{1.39}$$

for any $X, Y \in \Gamma(D^{\perp})$. Now, let $\bar{M}$ be a leaf of $D^{\perp}$. Denote by $\tilde{h}$ (resp. $\bar{h}$) the second fundamental form of the immersion of $\bar{M}$ in N (resp. M). Then we have

$$\tilde{h}(X, Y) = \bar{h}(X, Y) + h(X, Y), \tag{1.40}$$

for any $X, Y \in \Gamma(T\bar{M})$. Suppose $\bar{M}$ is totally geodesic in N. Then $\bar{M}$ is totally geodesic in M and by Theorem 1.4 and (1.40) we obtain (ii). Also, from (1.39) it follows that $C(\nabla^{\perp}_X \omega Y) = 0$ which is just (i).

Conversely, suppose (i) and (ii) are satisfied. Then from (1.39) we obtain $h(X, Y) = 0$ for all $X, Y \in \Gamma(D^{\perp})$. On the other hand, by (ii) and Theorem 1.4, $\bar{M}$ is totally geodesic in M. Hence by (1.40) we have $\tilde{h} = 0$, that is, M is totally geodesic in N.

§2. Semi-Invariant Products of Sasakian Manifolds

Let M be a semi-invariant submanifold of a Sasakian manifold N. We say that M is a semi-invariant product if the distribution $D \oplus \{\xi\}$ is integrable and locally M is a Riemannian product $M_1 \times M_2$, where M_1 (resp. M_2) is a leaf of $D \oplus \{\xi\}$ (resp. $D^\perp$). If we have $pq \neq 0$ we say that M is a <u>proper semi-invariant product</u>.

As is well known, Riemannian decomposable spaces are characterized by linear connections (see §4 of Chapter I and Yano [2] p. 219). We give first such a characterization for semi-invariant products.

THEOREM 2.1. <u>Let M be a semi-invariant submanifold of a Sasakian manifold N. Then M is a semi-invariant product if and only if</u>

$$\nabla_Y X \in \Gamma(D \oplus \{\xi\}), \quad \underline{\text{for any}} \quad Y \in \Gamma(TM) \quad \underline{\text{and}} \quad X \in \Gamma(D). \tag{2.1}$$

<u>Proof</u>. Suppose M is a semi-invariant product locally represented by $M_1 \times M_2$. Then M_1 and M_2 are totally geodesic in M and the Gauss formula implies

$$\nabla_Y X = \overset{(1)}{\nabla}_Y X \in \Gamma(D \oplus \{\xi\}) \quad \text{for any} \quad X, Y \in \Gamma(D \oplus \{\xi\}) \tag{2.2}$$

and

$$\nabla_V U = \overset{(2)}{\nabla}_V U \in \Gamma(D^\perp), \quad \text{for any} \quad U, V \in \Gamma(D^\perp), \tag{2.3}$$

where $\overset{(1)}{\nabla}$ amd $\overset{(2)}{\nabla}$ are the Riemannian connections on M_1 and M_2 respectively. On the other hand, by using (2.3) and (1.12) we obtain

$$\left.\begin{aligned} g(\nabla_V X, U) &= -g(X, \nabla_V U) = 0, \\ g(\nabla_\xi X, U) &= g(-\phi X + [\xi, X], U) = 0 \end{aligned}\right\} \tag{2.4}$$

for any $X \in \Gamma(D \oplus \{\xi\})$ and $U, V \in \Gamma(D^\perp)$. Thus from (2.2) and (2.4) it follows that (2.1) holds.

Conversely, suppose (2.1) is satisfied. Then the distribution $D \oplus \{\xi\}$ is integrable since we have

$$[Y, X] = \nabla_Y X - \nabla_X Y \in \Gamma(D \oplus \{\xi\}),$$

$$[X, \xi] = \nabla_X \xi - \nabla_\xi X = -\phi X - \nabla_\xi X \in \Gamma(D \oplus \{\xi\}),$$

for any $X, Y \in \Gamma(D)$. Moreover, if M_1 is a leaf of $D \oplus \{\xi\}$ then from (2.1) and the Gauss formula for the immersion of M_1 in M it follows that M_1 is totally geodesic in M. Finally, from (2.1) it follows that $\nabla_Y U \in \Gamma(D^\perp)$ for any $Y \in \Gamma(TM)$ and

$U \in \Gamma(D^{\perp})$. By using again the equation of Gauss for a leaf M_2 of $D^{\perp}$ we obtain that M_2 is totally geodesic in M. The proof is complete.

THEOREM 2.2 (Bejancu [13]). *Let M be a semi-invariant submanifold of a Sasakian manifold N. Then M is a semi-invariant product if and only if its second fundamental form satisfies*

$$Bh(X, Y) = 0 \tag{2.5}$$

or

$$h(X, \phi Y) = Ch(X, Y), \tag{2.6}$$

for any $X \in \Gamma(TM)$ and $Y \in \Gamma(D)$.

Proof. From (1.9) and (1.11) it follows that

$$\nabla_X \phi Y = \psi(\nabla_X Y) + Bh(X, Y) \tag{2.7}$$

and

$$h(X, \phi Y) = Ch(X, Y) + \omega(\nabla_X Y), \tag{2.8}$$

for any $X \in \Gamma(TM)$ and $Y \in \Gamma(D)$. Thus our assertion follows from (2.7) and (2.8) by means of Theorem 2.1.

Now, using the formulas of Gauss and Weingarten, we obtain

$$g(A_{\phi Z}X, Y) = -g(Bh(X, Y), Z), \tag{2.9}$$

for any $X \in \Gamma(D)$, $Y \in \Gamma(TM)$ and $Z \in \Gamma(D^{\perp})$. Then, by using (2.9) and Theorem 2.2, we get

COROLLARY 2.1. *Let M be a semi-invariant submanifold of a Sasakian manifold N. Then the following assertions are equivalent to each other:*

(i) *M is a semi-invariant product;*

(ii) *the fundamental tensors of Weingarten satisfy*

$$A_{\phi Z}X = 0, \tag{2.10}$$

for any $Z \in \Gamma(D^{\perp})$ and $X \in \Gamma(D)$;

(iii) *the second fundamental form of M satisfies*

$$h(Y, \phi X) = \phi h(Y, X), \tag{2.11}$$

for any $X \in \Gamma(D)$ and $Y \in \Gamma(TM)$.

Let $N(c)$ be a Sasakian space form of constant ϕ-sectional curvature c. Then the curvature tensor $\tilde{R}$ of $N(c)$ is given by (see §6 of Chapter I)

$$\tilde{R}(X, Y)Z = \frac{1}{4}(c + 3)\{g(Y, Z)X - g(X, Z)Y\} + \\ + \frac{1}{4}(c - 1)\{\eta(X)\cdot\eta(Z)Y - \eta(Y)\cdot\eta(Z)X + \\ + g(X, Z)\cdot\eta(Y)\xi - g(Y, Z)\cdot\eta(X)\xi + g(Z, \phi Y)\phi X - \\ - g(Z, \phi X)\phi Y + 2g(X, \phi Y)\phi Z\}, \tag{2.12}$$

for any X, Y, Z tangent to N(c). Suppose M is a semi-invariant submanifold of N(c). Then, by using (2.10) and (2.12), the equations of Codazzi and Ricci respectively become

$$(\nabla_X h)(Y, Z) - (\nabla_Y h)(X, Z) = \frac{1}{4}(c - 1) \times \\ \times \{g(Z, \psi Y)\omega X - g(Z, \psi X)\omega Y + 2g(X, \psi Y)\omega Z\} \tag{2.13}$$

and

$$\frac{1}{4}(c - 1)\{g(U, \phi Y)g(V, \phi X) - g(U, \phi X)g(V, \phi Y) + \\ + 2g(X, \phi Y)g(V, \phi U)\} = g(R^{\perp}(X, Y)U, V) + \\ + g([A_V, A_U]X, Y), \tag{2.14}$$

for any $X, Y, Z \in \Gamma(TM)$ and $U, V \in \Gamma(TM^{\perp})$, where $R^{\perp}$ is the curvature of the normal connection on M.

Now we shall prove the non-existence of proper semi-invariant products in some Sasakian space forms.

THEOREM 2.3 (Bejancu [13]). <u>There exist no proper semi-invariant products in Sasakian space forms N(c) with $c < -3$.</u>

<u>Proof</u>. Suppose M is a semi-invariant product in N(c). Then by (2.11) and (2.13) we obtain

$$\frac{1}{2}(1 - c)g(X, X)g(Z, Z) = g((\nabla_X h)(\phi X, Z), \phi Z) - \\ - g((\nabla_{\phi X} h)(X, Z), \phi Z) \tag{2.15}$$

for any $X \in \Gamma(D)$ and $Z \in \Gamma(D^{\perp})$. On the other hand, by using (2.1) we have

$$g((\nabla_X h)(\phi X, Z) - (\nabla_{\phi X} h)(X, Z), \phi Z) = \\ = g(\nabla^{\perp}_X h(\phi X, Z) - \nabla^{\perp}_X h(X, Z), \phi Z) + \\ + g(Bh(\nabla_X \phi X - \nabla_{\phi X} X, Z), Z). \tag{2.16}$$

Next, by using (2.11), (1.5), and (1.12) we get

$$g(\nabla^{\perp}_{X}h(\phi X,\ Z),\ \phi Z) = -g(h(X,\ Z),\ h(X,\ Z)) \qquad (2.17)$$

Replacing X by ϕX in (2.17) and taking account of (2.11) we have

$$g(\nabla^{\perp}_{\phi X}h(X,\ Z),\ \phi Z) = g(h(X,\ Z),\ h(X,\ Z)). \qquad (2.18)$$

Then by using (1.2), (1.5) and (1.12) we obtain

$$g(Bh(\nabla_{X}\phi X - \nabla_{\phi X}X,\ Z),\ Z) = 2g(X,\ X)g(Z,\ Z). \qquad (2.19)$$

Thus, from (2.16)-(2.19), it follows that

$$g((\nabla_{X}h)(\phi X,\ Z) - (\nabla_{\phi X}h)(X,\ Z),\ \phi Z) =$$

$$= 2\{g(X,\ X)g(Z,\ Z) - g(h(X,\ Z),\ h(X,\ Z))\}. \qquad (2.20)$$

Finally, (2.15) and (2.20) imply

$$\frac{1}{4}(c + 3)g(X,\ X)g(Z,\ Z) = g(h(X,\ Z),\ h(X,\ Z)). \qquad (2.21)$$

Since M is a proper semi-invariant product, from (2.21) it follows that $c \geqslant -3$. The proof is complete.

Let M be a semi-invariant product in N(c). If $q = \dim T_xM^{\perp}$ we say that M is a generic semi-invariant product. Then by using Theorem 2.3 we obtain

COROLLARY 2.2. There exist no proper generic semi-invariant products in a Sasakian space form N(c) with $c \neq -3$.

Proof. Suppose M is a proper generic semi-invariant product in N(c). Since we have $B = \phi$, (2.5) implies that $h(X,\ Z) = 0$ for any $X \in \Gamma(D)$ and $Z \in \Gamma(D^{\perp})$. Thus from (2.21) we have $c = -3$.

Let M be a semi-invariant submanifold of a Sasakian manifold N. Then we say that M is totally contact-umbilical if there exists a normal vector field H such that the second fundamental form of M is given by

$$h(X,\ Y) = g(\phi X,\ \phi Y)H + \eta(X)h(Y,\ \xi) + \eta(Y)h(X,\ \xi), \qquad (2.22)$$

for any vector fields X, Y tangent to M. If we have $H = 0$ in (2.22), that is the second fundamental form of M is given by

$$h(X,\ Y) = \eta(X)h(Y,\ \xi) + \eta(Y)h(X,\ \xi), \qquad (2.23)$$

then we say that M is totally contact-geodesic.

THEOREM 2.4. (Bejancu [13]). Any proper totally contact-umbilical semi-invariant submanifold of a Sasakian manifold with $q > 1$ is a totally contact-geodesic submanifold.

Proof. From (1.14) it follows that

$$A_{\phi X}(BH) = A_{\phi BH}(X), \tag{2.24}$$

for any $X \in \Gamma(D^{\perp})$. Since M is totally contact-umbilical, from (2.24) we obtain

$$g(X, X)g(BH, BH) = g(BH, X)^2. \tag{2.25}$$

From (2.25) we get $BH = 0$ since we supposed $q > 1$.

Next, by using $BH = 0$ and the formulas of Gauss and Weingarten we obtain

$$PA_{\phi H}Y = \psi(A_H Y), \tag{2.26}$$

for any $Y \in \Gamma(TM)$. We have further

$$g(PA_{\phi H}Y, Z) = g(A_{\phi H}Y, Z) = g(Y, Z)g(H, \phi H) = 0 \tag{2.27}$$

and

$$g(\psi(A_H Y), Z) = -g(A_H Y, \phi Z) = -g(Y, \phi Z)g(H, H), \tag{2.28}$$

for any $Y \in \Gamma(TM)$ and $Z \in \Gamma(D)$. Since M is a proper semi-invariant submanifold, (2.26)-(2.28) imply that $H = 0$. Hence M is totally contact-geodesic. The proof is complete.

THEOREM 2.5 (Bejancu [13]). Any totally contact-geodesic semi-invariant submanifold M of a Sasakian manifold N is locally a Riemannian product $M_1 \times M_2$ where M_1 is a totally geodesic invariant submanifold of N and M_2 is a totally geodesic anti-invariant submanifold of N and ξ is normal to M_2.

Proof. First, from (2.23) we obtain

$$h(X, Y) = 0, \quad \text{for any } X \in \Gamma(D) \text{ and } Y \in \Gamma(TM).$$

Thus, by Theorem 2.2, M is a semi-invariant product locally represented by $M_1 \times M_2$. On the other hand, by direct computation, using (2.23) and Corollary 1.1 we obtain that M_1 and M_2 are both totally geodesic in N.

From Theorems 2.4 and 2.5 the following holds

COROLLARY 2.3. Any proper totally contact-umbilical semi-invariant submanifold M of the Sasakian manifold N with $q > 1$ is locally the Riemannian product $M_1 \times M_2$ where M_1 and M_2 have the properties of Theorem 2.5.

§3. Semi-Invariant Submanifolds with Flat Normal Connection

Let S^{2n+1} be a real (2n+1)-dimensional sphere endowed with the standard Sasakian structure (see §6 of Chapter I). The main purpose of this paragraph is to give some results on the geometry of semi-invariant submanifolds with flat normal connection in S^{2n+1}.

First we have

LEMMA 3.1. Let M be a semi-invariant submanifold of S^{2n+1} with flat normal connection. Then $A_{CV} = 0$ for any vector field V normal to M.

LEMMA 3.2. Let M be a semi-invariant submanifold of S^{2n+1} with flat normal connection. If the mean curvature vector is parallel and if $\psi \circ A_V = A_V \circ \psi$ for any vector field V normal to M, then the second fundamental form of M is parallel.

Now we can state

THEOREM 3.1 (Yano-Kon [7]). Let M be an m-dimensional complete semi-invariant submanifold of S^{2n+1} with flat normal connection. If the mean curvature vector of M is parallel and if $\psi \circ A_V = A_V \circ \psi$ for any vector field normal to M, then M is an S^m or M is

$$S^{m_1}(r_1) \times \ldots \times S^{m_k}(r_k), \quad m = \sum_{i=1}^{k} m_i, \quad 2 \leqslant k \leqslant m,$$
$$\sum_{i=1}^{k} (r_i)^2 = 1,$$

in some S^{m+q}, where $m_1, \ldots, m_k$ are odd numbers.

Proof. First we assume $\omega = 0$, that is, M is an invariant submanifold of S^{2n+1}. Then the second fundamental form of M satisfies $\psi \circ A_V + A_V \circ \psi = 0$ (see Kon [2]). Thus by assumption we have $\psi \circ A_V = 0$ which implies $A_V X = 0$ for each $X \in \Gamma(D)$ since D is invariant by A_V and ψ is an automorphism of D. On the other hand, by using the Gauss formula, we obtain

$$g(A_V\xi, X) = g(h(X, \xi), V) = g(-\phi X, V) = 0,$$

for any $X \in \Gamma(D \oplus \{\xi\})$. Hence $A_V\xi = 0$. Therefore M is a totally geodesic submanifold of S^{2n+1}, that is, M is an S^m and m is odd.

Now we assume $\omega \neq 0$. Then it follows that $A_{\omega Y} \neq 0$ for each $Y \in \Gamma(TM)$. Hence by Lemma 3.1 the first normal space is of dimension q. It is easy to check that the product of spheres from the assertion is a semi-invariant submanifold with flat normal connection and parallel mean curvature vector. Then the assertion follows by using Theorem 2.2 of Chapter IV and Lemma 3.2.

From this theorem we have

COROLLARY 3.1. <u>Let M be a $(2p+q+1)$-dimensional complete submanifold of S^{2n+1} with flat normal connection and $q + p = n$. If the mean curvature vector of M is parallel and if $\psi\circ A_V = A_V\circ\psi$ for any vector field V normal to M, then M is</u>

$$S^{m_1}(r_1)\times\ldots\times S^{m_k}(r_k), \quad 2p + q + 1 = \sum_{i=1}^{k} m_i,$$
$$2 \leqslant k \leqslant 2p + q + 1, \quad \sum_{i=1}^{k} (r_i)^2 = 1,$$

<u>where $m_1,\ldots,m_k$ are odd numbers.</u>

Semi-invariant submanifolds with flat normal connection have also been studied in other Sasakian space forms. We note such a result here.

THEOREM 3.2 (M. Kobayashi [2]). <u>Let M be a $(2p+q+1)$-dimensional semi-invariant submanifold of a $(2n+1)$-dimensional Sasakian space form $N(c)$ with $c > 1$. Suppose M has flat normal connection. Then we have:</u>

(i) <u>if $2p + q \leqslant n$ then M is an anti-invariant submanifold;</u>

(ii) <u>if $2p + q > n$ then we have $n = p + q$.</u>

More results on the geometry of semi-invariant submanifolds (contact CR-submanifolds) of Sasakian manifolds can be found in Yano-Kon [7], Kobayashi [1]-[4], Hsu [1], Roşca [1], Papaghiuc [2], Bejancu-Papaghiuc [1], [2] and Bejancu [12], [13]. Taking into account the large variety of problems in the geometry of submanifolds, some other results are expected to be obtained.

§4. Generic submanifolds of Kaehlerian manifolds

The purpose of this section is to show that CR-submanifolds have an interesting generalization even in the case of almost Hermitian manifolds.

Let (N, J, g) be an almost Hermitian manifold and let M be a real submanifold of N. Then we say that M is a generic submanifold of N if the maximal complex subspaces $D_x = T_xM \cap J(T_xM)$ determine on M a distribution $D : x \to D_x \subset T_xM$. It is easy to prove that D is a differentiable distribution. The concept of generic submanifolds was first introduced by Chen in [9] for submanifolds in almost complex manifolds.

We denote by $D^\perp$ the orthogonal complementary distribution to D in TM and note that $JD^\perp \cap D^\perp = \{0\}$. If in particular, $JD^\perp \subset TM^\perp$ we have the concept of a CR-submanifold. When $D = \{0\}$ we say that M is a purely real submanifold of N. We call D and $D^\perp$ the holomorphic and purely real distributions on M respectively.

With respect to the geometry of generic submanifolds interesting results were obtained by Chen in [9] and Chen-Ludden-Montiel in [1]. On the other hand, the concept has also been considered for submanifolds in Sasakian manifolds by Verheyen [1] and Bejancu-Papaghiuc [3].

We shall sketch here some problems on the integrability of the distributions on a generic submanifold.

First, for each vector field X tangent to M we put

$$JX = \phi X + \omega X,$$

where ϕX and ωX are the tangential and normal components of JX, respectively. As in the case of CR-submanifolds, ω is a normal-bundle-valued 1-form of TM and ϕ is an endomorphism of TM.

Now we state

THEOREM 4.1 (Chen [9]). Let M be a generic submanifold of a Kaehler manifold N. Then we have:

(i) the holomorphic distribution D is integrable if and only if the second fundamental form of M satisfies

$$g(h(X, JY), JZ) = g(h(JX, Y), JZ),$$

for any $X, Y \in \Gamma(D)$ and $Z \in \Gamma(D^\perp)$;

(ii) the purely real distribution $D^\perp$ is integrable if and only if

$$\nabla_Z(\phi U) - \nabla_U(\phi Z) + A_{\omega Z}U - A_{\omega U}Z \in \Gamma(D^\perp),$$

for all $Z, U \in \Gamma(D^\perp)$.

Proof. The proof of (i) is analogous with that of Theorem 1.1 of Chapter III. For the second assertion, by using the equations of Gauss and Weingarten we get

$$J(\nabla_Z U) + Jh(Z, U) = \nabla_Z(\phi U) + h(Z, \phi U) -$$
$$- A_{\omega U}Z + \nabla_Z^\perp \omega U.$$

Thus we obtain

$$[Z, U] = \phi(\nabla_U(\phi Z) - \nabla_Z(\phi U) + A_{\omega U}Z - A_{\omega Z}U) +$$
$$+ B(\nabla_U^\perp \omega Z - \nabla_Z^\perp \omega U) + h(U, \phi Z) - h(Z, \phi U)$$

where BV is the tangent part of JV for a certain $V \in \Gamma(TM^\perp)$. Thus the assertion (ii) follows since in fact $BV \in \Gamma(D^\perp)$ for each $V \in \Gamma(TM^\perp)$.

Next, we say that a real sbumanifold M of a Kaehler manifold N is a generic product if it is locally the Riemannian product of a complex submanifold $M^\top$ and a purely real submanifold $M^\perp$ of N.

Of course, if we take a complex submanifold M of C^p and a purely real submanifold $M^\top$ of C^q, then $M^\top \times M^\perp$ is a generic product in C^{p+q}. Chen obtained in [9] examples of generic products in CP^n. On the other hand, there exist no proper generic products in N(c) with $c < 0$, that is, in a such space a generic product is either a complex manifold or a purely real submanifold. Finally, we note that Theorems 4.1 and 4.2 of Chapter IV and Theorem 5.2 of Chapter III were generalized by Chen-Ludden-Montiel [1] to the case of generic products.

§5. QR-Submanifolds of Quaternion Kaehlerian Manifolds

Let N be a 4n-dimensional manifold and g be a Riemannian metric on N. Then N is said to be a quaternion Kaehlerian manifold (see §7 of Chapter I), if there exists a 3-dimensional vector bundle V of tensors of type (1,1) with local basis of almost Hermitian structures J_1, J_2, J_3 satisfying

$$J_1 \circ J_2 = -J_2 \circ J_1 = J_3 \tag{5.1}$$

and

$$\tilde{\nabla}_X J_a = \sum_{b=1}^{3} Q_{ab}(X) J_b, \quad (a = 1, 2, 3), \tag{5.2}$$

for all vector fields X tangent to N, where $\tilde{\nabla}$ is the Levi-Civita connection determined by g on N and Q_{ab} are certain 1-forms locally defined on N such that $Q_{ab} + Q_{ba} = 0$.

Suppose $\{J_1, J_2, J_3\}$ is a local basis for the vector bundle V in a coordinate neighbourhood $\mathcal{U}$ on N. We take another coordinate neighbourhood $\tilde{\mathcal{U}}$ with the corresponding local basis $\{\tilde{J}_1, \tilde{J}_2, \tilde{J}_3\}$ and $\mathcal{U} \cap \tilde{\mathcal{U}} \neq \emptyset$. Then we have

$$\tilde{J}_a = \sum_{b=1}^{3} S_{ab} J_b, \quad (a = 1, 2, 3), \tag{5.3}$$

where $[S_{ab}]$ is an element of the special orthogonal group SO(3).

Now, let M be an m-dimensional Riemannian manifold isometrically immersed in N. The geometry of M depends on the behaviour of the tangent space to M under the action of the local basis $\{J_1, J_2, J_3\}$. Then we say that M is a <u>quaternion-real submanifold</u> (<u>QR-submanifold</u>) if there exists a vector subbundle ν of the normal bundle $TM^{\perp}$ such that we have

$$J_a(\nu_x) = \nu_x \tag{5.4}$$

and

$$J_a(\nu_x^{\perp}) \subset T_x M, \tag{5.5}$$

for each $x \in M$ and $a = 1, 2, 3$ where $\nu^{\perp}$ is the complementary orthogonal subbundle to ν in $TM^{\perp}$.

Of course, each quaternion submanifold (see Chen [2]) is a QR-submanifold with $\nu = TM^{\perp}$ and $\nu^{\perp} = \{0\}$. Moreover we have

PROPOSITION 5.1. <u>Let M be a real hypersurface of a quaternion Kaehlerian manifold N. Then M is a QR-submanifold.</u>

<u>Proof</u>. Since $T_x M^{\perp}$ is 1-dimensional and J_a are Hermitian structures we obtain

$$J_a(T_x M^{\perp}) \subset T_x M, \quad \text{for each} \quad x \in M.$$

Hence M is a QR-submanifold with $\nu = \{0\}$ and $\nu^{\perp} = TM^{\perp}$.

Also, each anti-quaternion submanifold of N (see Pak [2] and Bejancu [22]) is an example of a QR-submanifold.

Let M be a QR-submanifold of N. Then we denote $D_{ax} = J_a(\nu_x^{\perp})$ and remark that D_{1x}, D_{2x}, D_{3x} are mutually orthogonal vector subspaces of T_xM. We consider

$$D_x^{\perp} = D_{1x} \oplus D_{2x} \oplus D_{3x}$$

and by (5.3) obtain a 3s-dimensional distribution $D^{\perp} : x \to D_x^{\perp}$ globally defined on M, where $s = \dim \nu_x^{\perp}$. Also we have

$$J_a(D_{ax}) = \nu_x^{\perp}; \quad J_a(D_{bx}) = D_{cx}, \tag{5.6}$$

for each $x \in M$, where (a, b, c) is a cyclic permutation of (1, 2, 3). Next, we denote by D the complementary orthogonal distribution to $D^{\perp}$ in TM and obtain that D is an invariant distribution on M, i.e., we have

$$J_a(D_x) = D_x, \quad \text{for any} \quad x \in M, \quad a = 1, 2, 3. \tag{5.7}$$

We call D the quaternion distribution on M.

Now denote by P the projection morphism of TM to the quaternion distribution D and choose a local field of orthonormal frames $\{V_1,\ldots,V_s\}$ on the vector subbundle $\nu^{\perp}$ in $TM^{\perp}$. Then on the distribution $D^{\perp}$ we have the field of orthonormal frames

$$\{E_{11},\ldots,E_{1s}, E_{21},\ldots,E_{2s}, E_{31},\ldots,E_{3s}\}, \tag{5.8}$$

where $E_{ai} = J_aV_i$, a = 1, 2, 3 and i = 1,...,s. Thus any vector field Y tangent to M can be written locally as follows

$$Y = PY + \sum_{b=1}^{3} \sum_{i=1}^{s} \omega_{bi}(Y)E_{bi}, \tag{5.9}$$

where ω_{bi} are 1-forms locally defined on M by

$$\omega_{bi}(Y) = g(Y, E_{bi}).$$

Applying J_a to (5.9) and taking account of (5.1) we obtain

$$J_aY = J_aPY + \sum_{i=1}^{s} \{\omega_{bi}(Y)E_{ci} - \omega_{ci}(Y)E_{bi} - \omega_{ai}(Y)V_i\}. \tag{5.10}$$

On the other hand, for any normal section V we put

$$J_aV = B_aV + C_aV, \quad a = 1, 2, 3, \tag{5.11}$$

where B_aV and C_aV are respectively the tangent part and the normal part of J_aV. By using (5.10), (5.11) and equations of Gauss and Weingarten in (5.2) and then taking the normal parts we obtain

$$h(X, J_aPY) - C_ah(X, Y) + \sum_{i=1}^{s} \{\omega_{bi}(Y)h(X, E_{ci}) - $$
$$- \omega_{ci}(Y)h(X, E_{bi}) - \omega_{ai}(Y)\nabla^{\perp}_X V_i - X(\omega_{ai}(Y))V_i +$$
$$+ \omega_{ai}(\nabla_X Y)V_i + Q_{ab}(X)\omega_{bi}(Y)V_i +$$
$$+ Q_{ac}(X)\omega_{ci}(Y)V_i\} = 0, \tag{5.12}$$

for any X, Y tangent to M, with usual notations for geometrical objects on a submanifold (see §3 of Chapter I). Next, from (5.12) we obtain

$$h(X, J_aY) - C_ah(X, Y) + \sum_{i=1}^{s} \{\omega_{ai}(\nabla_X Y)V_i\} = 0, \tag{5.13}$$

for any $X, Y \in \Gamma(D)$.

We say that M is a <u>D-geodesic QR-submanifold</u> if we have

$$h(X, Y) = 0, \quad \text{for any} \quad X, Y \in \Gamma(D). \tag{5.14}$$

Then we can state

THEOREM 5.1. <u>Let M be a QR-submanifold of the quaternion Kaehlerian manifold N. Then the following assertions are equivalent</u>:

(i) <u>the second fundamental form of M satisfies</u>

$$h(X, J_aY) = h(Y, J_aX), \quad \underline{\text{for any}} \quad X, Y \in \Gamma(D); \tag{5.15}$$

(ii) <u>M is D-geodesic</u>;

(iii) <u>the distribution D is integrable</u>.

<u>Proof</u>. (i) ⇒ (ii). By (5.15) and (5.1) we obtain

$$h(J_3X, Y) = h(X, J_3Y) = h(X, (J_1 \circ J_2)Y) =$$
$$= h(J_1X, J_2Y) = h((J_2 \circ J_1)X, Y) = -h(J_3X, Y). \tag{5.16}$$

Thus, from (5.16) it follows that (5.14) holds since J_3 is an automorphism of D.

(ii) ⇒ (iii). Taking account of (5.14) in (5.13) we obtain

$$g(\nabla_X Y,\ E_{ai}) = 0 \quad \text{for any} \quad X,\ Y \in \Gamma(D),$$

$$a = 1,\ 2,\ 3 \quad \text{and} \quad i = 1,\dots,s.$$

Thus D is integrable and each leaf of D is totally geodesic in M.

(iii) ⇒ (i). Suppose D is integrable. Then we have

$$\omega_{ai}([X,\ Y]) = 0, \quad \text{for all} \quad X,\ Y \in \Gamma(D),$$

$$a = 1,\ 2,\ 3; \quad i = 1,\dots,s.$$

Thus from (5.13) taking into account that ∇ is a torsion-free connection we obtain (5.15). The proof is complete.

We denote by A_i the fundamental tensor of Weingarten with respect to V_i. Then we have

LEMMA 5.1. <u>Let M be a QR-submanifold of a quaternion Kaehlerian manifold N. Then</u>

$$A_i E_{aj} = A_j E_{ai}, \quad a = 1,\ 2,\ 3; \quad i,\ j = 1,\dots,s. \tag{5.17}$$

<u>Proof</u>. First, from (5.2) we obtain

$$\tilde{\nabla}_X J_a Z = J_a(\tilde{\nabla}_X Z) + Q_{ab}(X) J_b Z + Q_{ac}(X) J_c Z, \tag{5.18}$$

for any X, Z tangent to N, where (a, b, c) is a cyclic permutation of (1, 2, 3). Next, we take X tangent to M and replace Z by E_{ai}. By using the formulas of Gauss and Weingarten, (5.18) becomes

$$h(X,\ E_{ai}) = J_a(\nabla^{\perp}_X V_i) - J_a(A_i X) - \nabla_X E_{ai} +$$

$$+ Q_{ab}(X) E_{bi} + Q_{ac}(X) E_{ci}. \tag{5.19}$$

Finally, by (5.19) and (3.3) of Chapter I we obtain

$$g(A_j E_{ai},\ X) = g(h(X,\ E_{ai}),\ V_j) = g(J_a(\nabla_X V_i) -$$

$$- J_a(A_i X),\ V_j) = - g(\nabla^{\perp}_X V_i,\ E_{ai}) +$$

$$+ g(A_i X,\ E_{aj}) = g(A_i E_{aj},\ X),$$

for any X tangent to M. Thus we have the assertion of the lemma.

COROLLARY 5.1. *Let M be a QR-submanifold of a quaternion Kaehlerian manifold N. Then*

$$g(\nabla_{E_{ai}} E_{aj}, X) = g(\nabla_{E_{aj}} E_{ai}, X), \tag{5.20}$$

for any $X \in \Gamma(D)$ and $a = 1, 2, 3$.

Proof. By using (5.2) and the equation of Gauss we obtain

$$g(\nabla_{E_{ai}} E_{aj}, X) = g(\tilde{\nabla}_{E_{ai}} J_a V_j, X) = g(J_a(\tilde{\nabla}_{E_{ai}} V_j) +$$

$$+ Q_{ab}(X)E_{bj} + Q_{ac}(X)E_{cj}, X) = -g(\tilde{\nabla}_{E_{ai}} V_j, J_a X) =$$

$$= g(A_j E_{ai}, J_a X).$$

Thus by Lemma 5.1 we obtain the assertion.

Now we define differential 1-forms B_{aij} by

$$B_{aij}(X) = g(\nabla_{E_{ai}} E_{aj}, X), \quad a = 1, 2, 3;$$
$$i, j = 1,\ldots,s, \tag{5.21}$$

for any $X \in \Gamma(TM)$ and state

THEOREM 5.2. *Let M be a QR-submanifold of a quaternion Kaehlerian manifold N. Then the following assertions are equivalent:*

(i) *the distribution $D^{\perp}$ is integrable;*

(ii) *$B_{aij}(X) = 0$ for all $a = 1, 2, 3$; $i, j = 1,\ldots,s$; $X \in \Gamma(D)$;*

(iii) *$h(D, D^{\perp}) \subset \nu$.*

Proof. First, by the definition of a QR-submanifold we have

$$E_{ai} = J_b E_{ci} = - J_c E_{bi}, \tag{5.22}$$

for $i = 1,\ldots,s$ and (a, b, c) a cyclic permutation of $(1, 2, 3)$. By using (5.2), (5.20) and (5.22) we obtain

$$g([E_{ai}, E_{bj}], J_c X) = B_{aij}(X) + B_{bij}(X), \tag{5.23}$$

for any $X \in \Gamma(D)$. On the other hand, by using (5.17), (5.2) and the Weingarten equation we infer

$$g([E_{ai}, E_{aj}], X) = 0. \tag{5.24}$$

Thus by (5.23) and (5.24) we get the equivalence of (i) and (ii). Next, by using (5.2) we get

$$B_{aij}(X) = g(h(J_a X, E_{ai}), V_j). \tag{5.25}$$

Hence the equivalence of (ii) and (iii) follows from (5.25). The proof is complete.

From this theorem we obtain

COROLLARY 5.2. <u>Let M be a real hypersurface of a quaternion Kaehlerian manifold N. Then the distribution $D^\perp$ is integrable if and only if we have $h(D, D^\perp) = \{0\}$.</u>

Now, we say that a QR-submanifold of N is $D^\perp$ -geodesic if we have $h(D^\perp, D^\perp) = \{0\}$. Then with respect to the geometry of the leaves of D and $D^\perp$ we have

THEOREM 5.3. <u>Let M be a QR-submanifold of a quaternion Kaehlerian manifold N. Then we have the following assertions:</u>

(i) <u>if the quaternion distribution D is integrable, then each leaf of D is totally geodesic in N;</u>

(ii) <u>if the distribution $D^\perp$ is integrable then we have:</u>

(a) <u>each leaf of $D^\perp$ is totally geodesic in M;</u>

(b) <u>a leaf of $D^\perp$ is totally geodesic in N if and only if M is $D^\perp$-geodesic.</u>

<u>Proof</u>. Suppose D is integrable. Then since D is a quaternion distribution, each leaf of D is a quaternion submanifold of N. Thus the assertion (i) follows from Lemma 4 in Chen [2].

Now, we assume $D^\perp$ be integrable. Then by using (5.2) and Theorem 5.2 we obtain

$$g(\nabla_X E_{ai}, Z) = g(\tilde{\nabla}_X E_{ai}, Z) = g(J_a(\tilde{\nabla}_X V_i), Z) =$$

$$g(A_i X, J_a Z) = g(h(X, J_a Z), V_i) = 0,$$

$$a = 1, 2, 3; \quad i = 1, \ldots, s,$$

for any $X \in \Gamma(D^\perp)$ and $Z \in \Gamma(D)$. Thus we have

$$g(\nabla_X Y, Z) = 0, \quad \text{for any } X, Y \in \Gamma(D^\perp) \text{ and } Z \in \Gamma(D), \tag{5.26}$$

that is, each leaf of $D^\perp$ is totally geodesic in M. Next, from (5.26) it follows that

$$g(\tilde{\nabla}_X Y,\ Z) = 0, \quad \text{for any} \quad X,\ Y \in \Gamma(D^{\perp}) \text{ and } Z \in \Gamma(D). \tag{5.27}$$

On the other hand, we have

$$g(\tilde{\nabla}_X Y,\ V) = g(h(X,\ Y),\ V), \tag{5.28}$$

for any X, $Y \in \Gamma(D^{\perp})$ and $V \in \Gamma(TM^{\perp})$. Let M^* be a leaf of $D^{\perp}$ and h^* the second fundamental form of the immersion of M^* in N. Then, from (5.27) and (5.28) we have

$$g(h^*(X,\ Y),\ Z) = 0 \tag{5.29}$$

and

$$g(h^*(X,\ Y),\ V) = g(h(X,\ Y),\ V). \tag{5.30}$$

Thus the assertion $(ii)_b$ follows from (5.29) and (5.30).

By means of Theorems 5.1-5.3 we obtain

THEOREM 5.4. <u>Let M be a QR-submanifold of a quaternion Kaehlerian manifold N. Then M is locally a Riemannian product $M^{\top} \times M^{\perp}$, where $M^{\top}$ and $M^{\perp}$ are leaves of D and $D^{\perp}$ respectively, if and only if the second fundamental form of M satisfies</u>

$$h(D,\ D) = \{0\} \tag{5.31}$$

<u>and</u>

$$h(D,\ D^{\perp}) \subset \nu. \tag{5.32}$$

From this theorem we infer

COROLLARY 5.3. <u>Let M be a real hypersurface of a quaternion Kaehlerian manifold N. Then M is locally a Riemannian product $M^{\top} \times M^{\perp}$, where $M^{\top}$ and $M^{\perp}$ are leaves of D and $D^{\perp}$ respectively, if and only if the second fundamental form of M satisfies $h(D,\ TM) = \{0\}$.</u>

§6. <u>Totally Umbilical and Totally Geodesic QR-Submanifolds of a Quaternion Kaehlerian Manifold.</u>

Let M be a totally umbilical QR-submanifold of a quaternion Kaehlerian manifold, that is, we have

$$h(X,\ Y) = g(X,\ Y)H, \tag{6.1}$$

for any X, Y tangent to M, where H is the mean curvature vector defined by $H = (1/m)$ Trace (h) (see §3 of Chapter I).

THEOREM 6.1. *Let M be a totally umbilical QR-submanifold of a quaternion Kaehlerian manifold N. Then we have the following assertions:*

(i) the distribution $D^{\perp}$ is involutive;

(ii) the distribution D is involutive if and only if M is totally geodesic.

Proof. These follow by using (6.1) and Theorems 5.1 and 5.2.

From Theorem 5.4 we have a complete characterization of totally geodesic QR-submanifolds.

THEOREM 6.2. *Each totally geodesic QR-submanifold of a quaternion Kaehlerian manifold is locally a Riemannian product $M^{\top} \times M^{\perp}$ where $M^{\top}$ and $M^{\perp}$ are leaves of D and $D^{\perp}$ respectively.*

Now we state

THEOREM 6.3. *Let M be a totally umbilical QR-submanifold of a quaternion Kaehlerian manifold N. If we have dim $\nu_x^{\perp} > 1$ then M is totally geodesic.*

The proof is similar to that of Theorem 2.1 from Chapter III so we omit it here.

The remaining part of this section is devoted to the study of the existence of totally umbilical QR-submanifolds in curved quaternion Kaehlerian manifolds and of curved QR-submanifolds.

First we state

LEMMA 6.1. *Let M be a totally umbilical QR-submanifold of a quaternion manifold N. Then*

$$g(\tilde{R}(X, E_{ai})X, E_{ai}) + g(\tilde{R}(X, E_{ai})J_aX, V_i) = 0, \quad (6.2)$$

for any $X \in \Gamma(D)$, a = 1, 2, 3 and i = 1,...,s, where $\tilde{R}$ is the curvature tensor of N.

Proof. It suffices to prove (6.2) for a = 1. By using (5.2) and (5.22) we have

$$g(\tilde{\nabla}_X\tilde{\nabla}_{E_{1i}}J_1X,\ V_i) = -\ g(\tilde{\nabla}_X\tilde{\nabla}_{E_{1i}}J_1X,\ J_1E_{1i}) =$$

$$-\ g(\tilde{\nabla}_X\tilde{\nabla}_{E_{1i}}X,\ V_i) + Q_{12}(E_{1i})g(\tilde{\nabla}_XX,\ E_{2i}) +$$

$$+\ Q_{13}(X)g(\tilde{\nabla}_{E_{1i}}X,\ E_{3i}) + Q_{12}(X)g(\tilde{\nabla}_{E_{1i}}X,\ E_{2i})$$

$$+\ Q_{13}(E_{1i})g(\tilde{\nabla}_XX,\ E_{3i}),\quad \text{for any } X \in \Gamma(D). \qquad (6.3)$$

By Theorem 6.1 the distribution $D^{\perp}$ is involutive and thus by Theorem 5.3 any leaf of $D^{\perp}$ is totally geodesic in M. Thus we have

$$g(\tilde{\nabla}_{E_{1i}}X,\ E_{2i}) = -\ g(X,\ \nabla_{E_{1i}}E_{2i}) = 0$$

and

$$g(\tilde{\nabla}_{E_{1i}}X,\ E_{3i}) = -\ g(X,\ \nabla_{E_{1i}}E_{3i}) = 0.$$

On the other hand, taking account of (5.2) and (6.1) we obtain

$$g(\tilde{\nabla}_XX,\ E_{2i}) = -g(X,\ \tilde{\nabla}_XE_{2i}) = -\ g(X,\ J_2(\tilde{\nabla}_XV_i)) =$$

$$= -\ g(J_2X,\ A_iX) = -\ g(h(X,\ J_2X),\ V_i) = 0 \quad \text{and}$$

$$g(\tilde{\nabla}_XX,\ E_{3i}) = 0.$$

Hence (6.3) becomes

$$g(\tilde{\nabla}_X\tilde{\nabla}_{E_{1i}}J_1X,\ V_i) + g(\tilde{\nabla}_X\tilde{\nabla}_{E_{1i}}X,\ E_{1i}) = 0. \qquad (6.4)$$

In a similar way it follows that

$$g(\tilde{\nabla}_{E_{1i}}\tilde{\nabla}_XJ_1X,\ V_i) + g(\tilde{\nabla}_{E_{1i}}\tilde{\nabla}_XX,\ E_{1i}) = 0 \qquad (6.5)$$

and

$$g(\tilde{\nabla}_{[X,\ E_{1i}]}J_1X,\ V_i) + g(\tilde{\nabla}_{[X,\ E_{1i}]}X,\ E_{1i}) = 0. \qquad (6.6)$$

Thus the assertion of the lemma follows from (6.4)-(6.6) by using (1.4) of Chapter I.

Now we say that M is a proper QR-submanifold if $D \neq \{0\}$ and $\nu^{\perp} \neq \{0\}$. Then we have

THEOREM 6.4. <u>There exist no proper totally umbilical QR-submanifolds in positively or negatively curved</u>

quaternion Kaehlerian manifolds.

Proof. Suppose M is a proper totally umbilical QR-submanifold of N with $K_N \neq 0$. Taking account of (6.1) and of the Codazzi equation we obtain

$$g(\tilde{R}(X, Y)Z, V) = g(Y, Z)g(\nabla^{\perp}_X H, V) - - g(X, Z)g(\nabla^{\perp}_Y H, V), \tag{6.7}$$

for any X, Y, Z tangent to M and V normal to M. Now we take $X \in \Gamma(D)$, $Z = J_1 X$, $Y = E_{1i}$ and $V = V_i$ in (6.7) and obtain

$$g(\tilde{R}(X, E_{1i})J_1 X, V_i) = 0. \tag{6.8}$$

By using (6.2) and (6.8) we obtain $K_N(X \wedge E_{1i}) = 0$ which is a contradiction. The proof is complete.

PROPOSITION 6.1. Let M be a totally umbilical QR-submanifold of a quaternion Kaehlerian manifold N. Then

$$K_M(X \wedge E_{ai}) = \|H\|^2, \quad a = 1, 2, 3; \quad i = 1, \ldots, s, \tag{6.9}$$

for any unit vector X from the quaternion distribution D.

Proof. By the structure equation of Gauss (see (3.6) of Chapter I) we have

$$K_M(X \wedge Y) = K_N(X \wedge Y) + \|H\|^2, \tag{6.10}$$

for any orthonormal vectors X and Y tangent to M. Then we put $X \in \Gamma(D)$, $Y = E_{ai}$ in (6.10) and taking into account that $K_N(X \wedge E_{ai}) = 0$ obtain (6.9).

From Proposition 6.1 we have

COROLLARY 6.1. There exist no proper totally umbilical negatively curved QR-submanifolds in quaternion Kaehlerian manifolds.

COROLLARY 6.2. Any proper totally geodesic QR-submanifold of constant sectional curvature is flat.

Now, suppose M is an anti-quaternion manifold, i.e., M is a QR-submanifold with $\nu = \{0\}$. Then we say that M is a proper anti-quaternion product if both distributions D and $D^{\perp}$ are integrable, $D \neq \{0\}$ and M is locally a Riemannian product

$M^{\top} \times M^{\perp}$ where $M^{\top}$ is a leaf of D and $M^{\perp}$ is a leaf of $D^{\perp}$.

THEOREM 6.5. Let M be a proper anti-quaternion product of a quaternion space form N(c). Then c = 0.

Proof. By using Theorem 5.4 we obtain $h(D, TM) = \{0\}$. Hence we get

$$(\nabla_X h)(Y, Z) = \nabla_X^{\perp} h(Y, Z) - h(\nabla_X Y, Z) - h(Y, \nabla_X Z) = 0$$

for any $X, Y \in \Gamma(TM)$ and $Z \in \Gamma(D)$ since the quaternion distribution D is parallel with respect to the Levi-Civita connection on M. Thus the Codazzi equation for the immersion of M in N(c) becomes

$$[\tilde{R}(X, Y)Z]^{\perp} = 0. \tag{6.11}$$

Taking account of the special form of the curvature tensor field of N(c) (see §7 of Chapter I) and of (6.11) we get

$$\frac{c}{4}\sum_{a=1}^{3}\{g(J_aY, Z)J_aX - g(J_aX, Z)J_aY\} = 0. \tag{6.12}$$

Now we choose $Y \in \Gamma(D)$ and $X \in \Gamma(D^{\perp})$. Then from (6.12) we have either $c = 0$ or

$$\sum_{a=1}^{3}\{g(J_aY, Z)J_aX\} = 0. \tag{6.13}$$

We show that (6.13) leads to a contradiction. Since J_1X, J_2X and J_3X are mutually orthogonal, from (6.13) we get $g(J_aY, Z) = 0$, for $a = 1, 2, 3$. Finally, we take $Z = J_1Y$ and obtain $D = \{0\}$ which is a contradiction.

COROLLARY 6.3. There exist no anti-quaternion products in a quaternion space form N(c) with $c \neq 0$.

M. Barros, B.Y. Chen and F. Urbano [1] have introduced the notion of quaternion CR-submanifold of a quaternion Kaehlerian manifold as a natural generalization of the concept of CR-submanifold of a Kaehlerian manifold. More precisely, we say that M is a quaternion CR-submanifold of N if M is endowed with two differentiable orthogonal distributions D and $D^{\perp}$ such that

$$TM = D \oplus D^{\perp}; \; J_a(D_x) = D_x \quad \text{and} \quad J_a(D_x^{\perp}) \subset T_xM^{\perp},$$

for each $x \in M$ and $a = 1, 2, 3$. If $D^{\perp} = \{0\}$ (resp. $D = \{0\}$) a quaternion CR-submanifold becomes a quaternion submanifold (resp. totally real submanifold). On the other hand, the concept of generic submanifold has been also considered in the quaternion case by Martinez and Santos [1].

Remark 6.1. Between the two classes of QR-submanifolds and quaternion CR-submanifolds there exists no inclusion relation because a real hypersurface is a QR-submanifold and it is not a quaternion CR-submanifold and a totally real submanifold is a quaternion CR-submanifold and it is not a QR-submanifold. On the other hand, quaternion submanifolds lie in the intersection of the above classes.

We conclude this chapter by saying that the concept of CR-submanifold has been also considered in: (i) locally product spaces (see Matsumoto [1] and Bejancu [21]), (ii) para-Sasakian manifolds (see Ianus-Mihai [1]), (iii) locally conformal Kaehler manifolds (see Matsumoto [3]), (iv) Kenmotsu manifolds (see Papaghiuc [1]), (v) f-structures with complemented frames (see Mihai [1] and Ornea [1]), (vi) co-Kaehler manifolds (see Bejancu-Smaranda [1]). Also we note that Tashiro and Kim [1] starting from the structures induced on a submanifold of a Kaehlerian manifold introduced and studied metric compound structures on Riemannian manifolds.

Chapter VI

CR-STRUCTURES AND PSEUDO-CONFORMAL MAPPINGS

§1. CR-Manifolds and f-Structures with Complemented Frames

CR-manifolds were intensively studied from the analytic point of view (see Wells [1], [2]). As is well known, complex manifolds and normal almost contact manifolds (see Blair [3], p. 62) are examples of CR-manifolds. Non-trivial CR-manifolds appeared as boundaries of domains in complex spaces, which in fact are real hypersurfaces (i.e., particular CR-submanifolds).

On the other hand, in 1963, Yano introduced in [1] the concept of manifold endowed with an f-structure which is a generalization of both a complex manifold and an almost contact manifold. Important results on the geometry of f-structures with complemented frames have been obtained by Goldberg-Yano [1], [2], Blair [1], [2], Blair-Ludden-Yano [1].

It is the purpose of this section to show the relationship between CR-manifolds and f-structures with complemented frames. Then we shall apply this relationship to the study of submanifolds of complex manifolds and pseudo-conformal mappings.

Let M be a $(2n + s)$-dimensional real differentiable manifold and D be a differentiable distribution on M of real dimension 2n. Suppose D is endowed with a morphism $J : D \to D$ of vector bundles satisfying $J^2 = -I$, where I is the identity morphism on D. Then we say that M is endowed with an almost complex distribution (D, J). If the following conditions are fulfilled:

$$[JX, JY] - [X, Y] \in \Gamma(D), \tag{1.1}$$

$$[J, J](X, Y) = [JX, JY] - [X, Y] - J([X, JY] + [JX, Y]) = 0, \tag{1.2}$$

for all $X, Y \in \Gamma(D)$, then we say that (D, J) defines a real CR-structure on M.

Next, we denote by $T_C M$ the complexified tangent bundle to M. A complex CR-structure on M is a complex subbundle H of $T_C M$ satisfying conditions (see §1 of Chapter II)

(i) $H \cap \bar{H} = \{0\}$,

(ii) H is involutive, i.e., for any complex vector fields U and V in H, [U, V] is also in H.

THEOREM 1.1. A differentiable manifold has a real CR-structure if and only if it has a complex CR-structure.

Proof. Suppose M has a real CR-structure (D, J). Then we define

$$H = \{X - \sqrt{-1}\cdot JX;\ X \in \Gamma(D)\}. \tag{1.3}$$

Of course we have $H \cap \bar{H} = \{0\}$. Moreover, if we take $U = X - \sqrt{-1}\cdot JX$ and $V = Y - \sqrt{-1}\cdot JY$ from H we obtain

$$[U, V] = [X, Y] - [JX, JY] - \sqrt{-1}\cdot\{[X, JY] + [JX, Y]\}. \tag{1.4}$$

Taking account of (1.2), (1.4) becomes

$$[U, V] = [X, Y] - [JX, JY] - \sqrt{-1}\cdot J\{[X, Y] - [JX, JY]\}. \tag{1.5}$$

Thus, by using (1.1) and (1.3), we obtain that [U, V] belongs to H. Consequently M has a complex CR-structure.

Conversely, suppose M has a complex CR-structure. Then we define the distribution D by

$$D = \{X = \mathrm{Re}(U);\ U \in H\} \quad \text{and} \quad J : D \to D \tag{1.6}$$

given by

$$JX = \mathrm{Re}(\sqrt{-1}\cdot U), \quad \text{where} \quad X = \mathrm{Re}(U) \quad \text{and} \quad U \in H. \tag{1.7}$$

Then it is easy to check that we have $J^2 = -I$. On the other hand, by using (1.6) and (1.7) we get

$$[JX, JY] - [X, Y] = -\mathrm{Re}([U, V]) \in \Gamma(D), \tag{1.8}$$

where $X = \mathrm{Re}(U)$ and $Y = \mathrm{Re}(V)$, that is, condition (1.1) is satisfied. Substituting Y by JY in (1.8) we obtain

$$[JX, Y] + [X, JY] = \mathrm{Re}(\sqrt{-1}\cdot[U, V]) \in \Gamma(D). \tag{1.9}$$

The condition (1.2) follows from (1.8) and (1.9). The proof is complete.

By this theorem we can say that we have a CR-structure on M either when M has a real CR-structure or a complex CR-structure. A manifold endowed with a CR-structure is called a CR-manifold.

Now, suppose M is a real (2n+s)-dimensional manifold endowed with an f-structure ϕ, that is, there exists a tensor field ϕ of type (1, 1) and of rank 2n on M satisfying

$$\phi^3 + \phi = 0. \tag{1.10}$$

Besides, we suppose there exist s vector fields $\{\xi_1,\ldots,\xi_s\}$ and s 1-forms $\{\eta^1,\ldots,\eta^s\}$ satisfying

$$\left.\begin{array}{l}\phi^2 = -I + \sum_{a=1}^{s} \eta^a \otimes \xi_a;\ \phi\xi_a = 0,\\ \eta^a(\xi_b) = \delta^a_b;\ \eta^a\circ\phi = 0,\end{array}\right\} \tag{1.11}$$

where I is the identity morphism on TM. Then we say that M is endowed with an f-structure with complemented frames (see Goldberg-Yano [1]) given by tensor fields (ϕ, η^a, ξ_a). Denote by D the distribution defined by

$$D = \{X \in \Gamma(TM);\ \eta^a(X) = 0,\ a = 1,\ldots,s\},$$

and by using the first equality in (1.11) we obtain that the restriction J of ϕ to D is an almost complex structure on D. Thus we have an almost complex distribution (D, J) on M with dim D_x = 2n for each $x \in M$.

The torsion tensor field S of an f-structure with complemented frames (ϕ, η^a, ξ_a) is defined by

$$S = [\phi, \phi] + 2 \sum_{a=1}^{s} d\eta^a \otimes \xi_a \tag{1.12}$$

where $[\phi, \phi]$ is the Nijenhuis tensor field of ϕ. We say that (ϕ, η^a, ξ_a) is D-normal if the torsion tensor field S vanishes on D, i.e., we have

$$S(X, Y) = 0, \quad \text{for any} \quad X, Y \in \Gamma(D). \tag{1.13}$$

PROPOSITION 1.1. Let M be a manifold endowed with a D-normal f-structure with complemented frames. Then M is a CR-manifold.

Proof. As we have seen above we have an almost complex distribution (D, J) on M. By applying η^b to (1.13) and taking account of (1.11) and (1.12) we obtain

$$0 = \eta^b([\phi, \phi](X, Y) - \sum_{a=1}^{s} \eta^a([X, Y])\xi_a) =$$
$$= \eta^b([JX, JY] - [X, Y]), \quad \text{for any } X, Y \in \Gamma(D).$$

Thus (1.1) is satisfied and we have $[JX, Y] + [X, JY] \in \Gamma(D)$, for any $X, Y \in \Gamma(D)$. Finally, by using (1.11) and (1.12) in (1.13) we obtain (1.2). Hence M is a CR-manifold.

Let M be a (2n+s)-dimensional CR-manifold and (D, J) be the CR-structure on M with $2n = \dim_R D_x$, $x \in M$. We can choose s linearly independent vector fields $\{\xi_1,\ldots,\xi_s\}$ defined on a coordinate neighborhood $\mathcal{U}$ of M, such that each ξ_a does not belong to the distribution D. Then we define s differential 1-forms $\{\eta^1,\ldots,\eta^s\}$ on $\mathcal{U}$ by

$$\eta^a(X + \sum_{b=1}^{s} \alpha^b \xi_b) = \alpha^a, \tag{1.14}$$

where α^a (a = 1,...,s) are differentiable functions on $\mathcal{U}$ and $X \in \Gamma(D)$. It is easy to check that η^a are s linearly independent 1-forms on $\mathcal{U}$ defining locally the distribution D, i.e., we have $\eta^a(X) = 0$ for each $X \in \Gamma(D)$. Moreover, we have $\eta^a(\xi_b) = \delta^a_b$.

Next, for each vector field $X \in \Gamma(TM)$ the vector field

$$Y = X - \sum_{b=1}^{s} \eta^b(X)\xi_b$$

belongs to D since $\eta^a(Y) = 0$, a = 1,...,s. Thus a tensor field ϕ of type (1, 1) is well defined on $\mathcal{U}$ by

$$\phi X = J(X - \sum_{b=1}^{s} \eta^b(X)\xi_b), \quad \text{for any } X \in \Gamma(TM). \tag{1.15}$$

Finally, by a direct computation we can verify (1.11) and (1.13). Thus we obtain

PROPOSITION 1.2. <u>On each coordinate neighborhood $\mathcal{U}$ of a CR-manifold M there exists a D-normal f-structure with complemented frames.</u>

The f-structure (ϕ, η^a, ξ_a) is called an <u>associate f-structure with complemented frames</u> to the CR-structure (D, J) on $\mathcal{U}$. Now, suppose $(\tilde{\phi}, \tilde{\eta}^a, \tilde{\xi}_a)$ is another associate f-structure to the same CR-structure (D, J) on $\mathcal{U}$. Then we have

LEMMA 1.1. The associate f-structures with complemented frames (ϕ, η^a, ξ_a) and $(\tilde{\phi}, \tilde{\eta}^a, \tilde{\xi}_a)$ are related by

$$\left.\begin{aligned} \tilde{\xi}_a &= A_a + \sum_{b=1}^{s} \lambda_a^b \xi_b, \\ \tilde{\eta}^a &= \sum_{b=1}^{s} \tilde{\lambda}_b^a \eta^b, \quad a = 1,\ldots,s, \\ \tilde{\phi} &+ \sum_{a=1}^{s} \tilde{\eta}^a \otimes JA_a = \phi, \end{aligned}\right\} \tag{1.16}$$

on $\mathcal{U}$, where $[\lambda_a^b]$ is an invertible matrix of differentiable functions, $[\tilde{\lambda}_a^b]$ is the inverse matrix of $[\lambda_b^a]$ and $A_a \in \Gamma(D)$.

For the f-structure (ϕ, η^a, ξ_a) associate to the CR-structure (D, J) on $\mathcal{U}$ we define the tensor fields G^a $(a = 1,\ldots,s)$ of type $(0, 2)$ by

$$G^a(X, Y) = d\eta^a(\phi X, Y), \tag{1.17}$$

for any X, Y tangent to M on $\mathcal{U}$. Then we have

PROPOSITION 1.3. The tensor fields G^a are symmetric and Hermitian on the almost complex distribution D, i.e., we have

$$G^a(X, Y) = G^a(Y, X) \tag{1.18}$$

and

$$G^a(JX, JY) = G^a(X, Y), \quad \text{for all} \quad X, Y \in \Gamma(D). \tag{1.19}$$

Proof. By (1.17) we obtain

$$G^a(X, Y) = -\frac{1}{2}\eta^a([JX, Y]) \quad \text{for any} \quad X, Y \in \Gamma(D), \tag{1.20}$$

since the restriction of ϕ to D is just J and $\eta^a(JX) = \eta^a(Y) = 0$. On the other hand, from (1.1) it follows that

$$\eta^a([JX, Y] + [X, JY]) = 0, \quad \text{for any} \quad X, Y \in \Gamma(D). \tag{1.21}$$

Thus we obtain our assertion from (1.20) and (1.21).

Now we consider another f-structure $(\tilde{\phi}, \tilde{\eta}^a, \tilde{\xi}_a)$ associate to the CR-structure (D, J) and denote by $\tilde{G}^a$ the tensor fields defined by (1.17). Then by using (1.20) and the second equality in (1.16) we obtain

$$\tilde{G}^a(X, Y) = \sum_{b=1}^{s} \lambda_b^a G^b(X, Y), \quad \text{for any } X, Y \in \Gamma(D). \tag{1.22}$$

PROPOSITION 1.4. <u>Let M be a $(2n + s)$-dimensional CR-manifold and let (D, J) be the CR-structure on M with $2n = \dim_R D_x$, $x \in M$. Suppose for each $X \in \Gamma(D)$ there exists $a \in \{1,\ldots,s\}$ such that $G^a(X, X) \neq 0$. Then the distribution D is not integrable and the dimension of an integral submanifold of D is less than $n + 1$.</u>

<u>Proof</u>. By the hypothesis and (1.20) we obtain that D is not integrable. Now suppose M is an integral submanifold of D of dimension $n + 1$. Then there exists a local field of frames $\{X_1,\ldots,X_n, JX_1\}$ on M^* since otherwise it follows $\dim_R D_x > 2n$. But, there exists $a \in \{1,\ldots,s\}$ such that $G^a(X_1, X_1) \neq 0$, that is $\eta^a([JX_1, X_1]) \neq 0$. Thus we have $[JX_1, X_1] \notin \Gamma(D)$ which contradicts $[JX_1, X_1] \in \Gamma(TM^*)$.

From this proposition we have

COROLLARY 1.1. <u>Let M be a $(2n + 1)$-dimensional CR-manifold such that G given by (1.17) is positive or negative definite. Then the distribution D is not integrable and the dimension of an integral submanifold of D is less than $n + 1$.</u>

Now, let M be a $(2n + s)$-dimensional manifold endowed with an f-structure ϕ of rank $2n$. Then we define $P = -\phi^2$ and $Q = I + \phi^2$ and we have

$$P + Q = I, \quad P^2 = P, \quad Q^2 = Q, \quad PQ = QP = 0,$$

that is, P and Q are two complementary projection morphisms on TM. Thus we have two complementary distributions D and D defined by

$$D = \{X \in \Gamma(TM);\ QX = 0\}, \quad \tilde{D} = \{X \in \Gamma(TM);\ PX = 0\}. \tag{1.23}$$

We say that the f-structure ϕ is <u>D-normal</u> if we have

$$[\phi, \phi](X, Y) = Q([X, Y]), \quad \text{for any } X, Y \in \Gamma(D). \tag{1.24}$$

It is easy to check that each normal almost contact manifold (see Blair [3]) has a D-normal f-structure. Thus we have a justification for the above definition. On the other hand, we see that each f-structure with complemented frames (ϕ, η^a, ξ_a) satisfying (1.13) is also a D-normal f-structure in this sense.

Now we state

THEOREM 1.2. Any real manifold endowed with a D-normal f-structure is a CR-manifold.

Proof. Suppose M is endowed with a D-normal f-structure ϕ. Then we have the distribution D given by (1.23) and the restriction of ϕ to D is just an almost complex structure J on D. Thus M is endowed with an almost complex distribution (D, J). Moreover, (1.24) becomes

$$[JX, JY] - [X, Y] - \phi\{[X, JY] + [JX, Y]\} = 0, \quad (1.25)$$

for any $X, Y \in \Gamma(D)$. Now, from (1.25) taking into account the definition of the projection morphisms P and Q we obtain (1.1) since we have $Q\phi = \phi + \phi^3 = 0$. Then (1.25) becomes (1.2). Thus by Theorem 1.1 the proof is complete.

From this theorem we get

COROLLARY 1.2 (Blair [3]). Any normal almost contact manifold is a CR-manifold.

§2. Generic Submanifolds of Complex Manifolds

Let $\tilde{M}$ be a complex manifold of complex dimension p and let M be a submanifold of $\tilde{M}$ of real dimension m. Then, denoting by $\tilde{J}$ the almost complex structure on $\tilde{M}$, we have

$$[\tilde{J}, \tilde{J}](X, Y) = 0, \quad \text{for any} \quad X, Y \in \Gamma(T\tilde{M}). \quad (2.1)$$

Now let $D_x = T_xM \cap \tilde{J}(T_xM)$, $x \in M$ so that D_x is the maximal invariant subspace D_x of T_xM under the action of $\tilde{J}$. Then we say that M is a generic submanifold of $\tilde{M}$ if $D : x \to D_x \subset T_xM$ is a distribution on M. Chen proved in [8] that in fact D is a differentiable distribution.

We denote by J the restriction of $\tilde{J}$ to the distribution D. Then (D, J) is an almost complex distribution on M of real dimension 2n. Moreover, we have

THEOREM 2.1. Each generic submanifold of a complex manifold is a CR-manifold.

Proof. We show that (D, J) defined above is a CR-structure on M. First, we consider a coordinate neighborhood $\mathcal{U}$ on M and take a complementary distribution $\tilde{D}$ to D on $\mathcal{U}$. Denote by P and $\tilde{P}$ the projection morphisms of TM to D and respectively $\tilde{D}$. Then

$$X = PX + \tilde{P}X, \quad \text{for any} \quad X \in \Gamma(TM). \tag{2.2}$$

Next, we see that $\tilde{J}PX = JPX \in \Gamma(D)$ and $\tilde{J}\tilde{P}X$ is not tangent to M, otherwise D is not the maximal holomorphic distribution on M. Thus, by using (2.1) and (2.2), we obtain

$$[JX, JY] - [X, Y] - JP\{[JX, Y] + [X, JY]\} = 0 \tag{2.3}$$

and

$$\tilde{P}\{[JX, Y] + [X, JY]\} = 0, \quad \text{for any } X, Y \in \Gamma(D). \tag{2.4}$$

Replacing X by JX in (2.4) we obtain (1.1) and then (1.2) is a consequence of (2.3) taking account of (2.2) and (2.4). The proof is complete.

Let M be a (2n + s)-dimensional generic submanifold of the complex manifold $\tilde{M}$. Suppose (D, J) is the CR-structure stated by Theorem 2.1. By the theory in §1 on each coordinate neighborhood $\mathcal{U}$ on M there exists an associate f-structure with complemented frames (ϕ, η^a, ξ_a), where ξ_a does not belong to D and η^a are 1-forms defining locally the distribution D. Thus we obtain on $\mathcal{U}$ the vector fields $N_a = \tilde{J}\xi_a$, a = 1,...,s, which are not tangent to the generic submanifold M. For this reason we call them the affine normals induced by the f-structure (ϕ, η^a, ξ_a). If $(\tilde{\phi}, \tilde{\eta}^a, \tilde{\xi}_a)$ is another f-structure with complemented frames associated to (D, J) and $\tilde{N}_a$ are the corresponding affine normals, by (1.16) we have

$$\left.\begin{aligned} \tilde{\xi}_a &= A_a + \sum_{b=1}^{s} \lambda_a^b \xi_b \\ \tilde{N}_a &= JA_a + \sum_{b=1}^{s} \lambda_a^b N_b \\ \tilde{\eta}^a &= \sum_{b=1}^{s} \tilde{\lambda}_b^a \eta^b; \quad \tilde{\phi} + \sum_{b=1}^{s} \tilde{\eta}^b \otimes JA_b = \phi. \end{aligned}\right\} \tag{2.5}$$

If $s = \text{codim } M$ we say that M is an affine anti-holomorphic submanifold. It is interesting to note that in this case (2.5) is sufficient for a complete study of the geometry of M.

§3. Anti-Holomorphic Submanifolds of Complex Manifolds

As we have seen, the Riemannian metric on the ambient space was an important aspect to the definition of a CR-submanifold. It is the purpose of this section to define CR-submanifolds in complex manifolds, that is, without using a Riemannian metric, and to sketch the main lines of a study for anti-holomorphic submanifolds.

Let $\tilde{M}$ be a complex manifold of complex dimension p and M be a real submanifold of $\tilde{M}$ of real dimension m. Then we say that M is a CR-submanifold of $\tilde{M}$ if there exist two differentiable distributions D and $\tilde{D}$ on M satisfying

(i) $T_xM = D_x \oplus \tilde{D}_x$;

(ii) $\tilde{J}(D_x) = D_x$; $\tilde{J}(\tilde{D}_x) \cap T_xM = \{0\}$,

for any $x \in M$, where $\tilde{J}$ is the almost complex structure on $\tilde{M}$. A CR-submanifold M with $m = 2n + s$, where $s = \dim_R \tilde{D}_x$ and $2n = \dim_R D_x$ immersed in a complex manifold M with $p = n + s$ is called an anti-holomorphic submanifold.

From now on, in this section we suppose that M is an anti-holomorphic submanifold of a complex manifold $\tilde{M}$. Then by the definition, $\tilde{J}\tilde{D}$ is a vector bundle satisfying

$$T\tilde{M} = D \oplus \tilde{D} \oplus \tilde{J}\tilde{D}. \tag{3.1}$$

Hence $\tilde{J}\tilde{D}$ should be considered as the normal bundle to M.

In order to study the geometry of M we fix a symmetric affine connection $\tilde{\nabla}$ on $\tilde{M}$ such that we have

$$\tilde{\nabla}_X\tilde{J} = 0, \quad \text{for any } X \in \Gamma(T\tilde{M}). \tag{3.2}$$

For the existence of a such affine connection see Walker [1] or Yano [2]. Then we put

$$\tilde{\nabla}_XY = \nabla_XY + h(X, Y), \quad \text{for any } X, Y \in \Gamma(TM), \tag{3.3}$$

where $\nabla_XY \in \Gamma(TM)$ and $h(X, Y) \in \Gamma(\tilde{J}\tilde{D})$. Certainly, by using the properties of $\tilde{\nabla}$ we get that ∇ is a symmetric affine connection on M and h is a normal bundle-valued symmetric bilinear form on M. We call h the second fundamental form of M.

We denote by P and Q the projection morphisms of TM to D and $\tilde{D}$ respectively. Then we define the morphisms $\phi : TM \to TM$ and $\omega : TM \to \tilde{J}\tilde{D}$ by

$$\phi X = \tilde{J}PX \tag{3.4}$$

and

$$\omega X = \tilde{J}QX, \tag{3.5}$$

respectively, for any $X \in \Gamma(TM)$. Thus we have

$$\tilde{J}X = \phi X + \omega X. \tag{3.6}$$

Next, let $U \in \Gamma(\tilde{J}\tilde{D})$ and $X \in \Gamma(TM)$. Then we put

$$\tilde{\nabla}_X U = - A_U X + \nabla^{\perp}_X U, \tag{3.7}$$

where $A_U X \in \Gamma(TM)$ and $\nabla^{\perp}_X U \in \Gamma(\tilde{J}\tilde{D})$. By the properties of $\tilde{\nabla}$ it follows that A_U is an endomorphism of TM and $\nabla^{\perp}$ is an affine connection on $\tilde{J}\tilde{D}$.

THEOREM 3.1. <u>Let M be an anti-holomorphic submanifold of a complex manifold $\tilde{M}$. Then we have:</u>

(i) <u>the distribution D is integrable if and only if the second fundamental form h of M satisfies</u>

$$h(X, \tilde{J}Y) = h(Y, \tilde{J}X), \quad \underline{\text{for any}}\ X, Y \in \Gamma(D); \tag{3.8}$$

(ii) <u>the distribution $\tilde{D}$ is integrable if and only if</u>

$$A_{\tilde{J}V}W = A_{\tilde{J}W}V, \quad \underline{\text{for any}}\ V, W \in \Gamma(\tilde{D}). \tag{3.9}$$

<u>Proof</u>. We take $X, Y \in \Gamma(D)$ and by using (3.2), (3.3) and (3.6) we obtain

$$\nabla_X \tilde{J}Y + h(X, \tilde{J}Y) = \phi(\nabla_X Y) + \omega(\nabla_X Y) + \tilde{J}h(X, Y). \tag{3.10}$$

Since $h(X, Y) \in \Gamma(\tilde{J}\tilde{D})$ we infer that $\tilde{J}h(X, Y) \in \Gamma(\tilde{D})$. Thus, taking the component in $\tilde{J}\tilde{D}$ in (3.10), we obtain

$$h(X, \tilde{J}Y) = \omega(\nabla_X Y).$$

Then taking into account that ∇ is a torsion-free affine connection we get

$$h(X, \tilde{J}Y) - h(Y, \tilde{J}X) = \omega([X, Y]). \tag{3.11}$$

Now it is easy to see that assertion (i) follows from (3.11).

Next, by using (3.3) and (3.7) in (3.2) we obtain

$$\nabla^{\perp}_V \tilde{J}W = \omega(\nabla_V W), \quad \text{for any}\ V, W \in \Gamma(\tilde{D}).$$

Thus by again using (3.7) we get

$$\tilde{\nabla}_V \tilde{J}W - \tilde{\nabla}_W \tilde{J}V = A_{\tilde{J}V}W - A_{\tilde{J}W}V + \omega([V, W]). \qquad (3.12)$$

On the other hand, we have

$$\tilde{\nabla}_V \tilde{J}W - \tilde{\nabla}_W \tilde{J}V = \tilde{J}([V, W]). \qquad (3.13)$$

Hence by (3.12), (3.13), and (3.6) we infer that

$$\phi([V, W]) = A_{\tilde{J}V}W - A_{\tilde{J}W}V,$$

which proves the assertion (ii) since $V \in \Gamma(\tilde{D})$ if and only if $\phi V = 0$. The proof is complete.

By Theorem 2.1, each anti-holomorphic submanifold is a CR-manifold. In this case the torsion tensor S defined by (1.12) is given by

$$S(X, Y) = [\phi, \phi](X, Y) - 2\tilde{J}d\omega(X, Y), \qquad (3.14)$$

for any $X, Y \in \Gamma(TM)$, where we have

$$d\omega(X, Y) = \frac{1}{2}\{\nabla^{\perp}_X \omega Y - \nabla^{\perp}_Y \omega X - \omega([X, Y])\}.$$

Of course, by Proposition 1.2 the f-structure ϕ is D-normal. According to the terminology of Yano-Ishihara [3] this means D is a <u>torsionless distribution</u>. When S vanishes identically on M we say that M is a <u>normal anti-holomorphic submanifold</u>.

THEOREM 3.2. <u>An anti-holomorphic submanifold M of the complex manifold $\tilde{M}$ is normal if and only if</u>

$$A_{N_a} \circ \phi = \phi \circ A_{N_a}, \quad a = 1, \ldots, s,$$

<u>where N_a is the local basis in $\tilde{J}\tilde{D}$.</u>

This theorem should be considered as a version of Theorem 3.1 of Chapter III for the non-metrical case.

§4. Pseudo-Conformal Mappings

Let M and M' be two CR-manifolds of the same real dimension 2n + s. We denote by (D, J) and (D', J') the CR-structures on M and respectively M'. A diffeomorphism $\Phi : M \to M'$ is called a <u>pseudo-conformal mapping</u> if $\Phi_* X \in \Gamma(D')$ for each $X \in \Gamma(D)$ and $\Phi_* \circ J = J' \circ \Phi_*$ where Φ_* is the differential mapping of Φ.

Now, suppose $\Phi : M \to M'$ be a pseudo-conformal mapping. Let (ϕ, η^a, ξ_a) be an f-structure with complemented frames associated to the CR-structure (D, J) on a coordinate neighborhood $\mathcal{U}$ of M. Then $\mathcal{U}' = \Phi(\mathcal{U})$ is a coordinate neighborhood on M' on which we define

$$\tilde{\phi} = \Phi_* \circ \phi \circ \Phi_*^{-1}; \quad \tilde{\xi}_a = \Phi_* \xi_a \quad \text{and} \quad \tilde{\eta}^a = \eta^a \circ \Phi_*^{-1}. \tag{4.1}$$

It is easy to check that $(\tilde{\phi}, \tilde{\eta}^a, \tilde{\xi}_a)$ is an f-structure with complemented frames associated to (D', J') on $\mathcal{U}'$. We take another f-structure $(\phi', \eta^{a'}, \xi'_a)$ associated to (D', J') on $\mathcal{U}'$ and by using (4.1) and (1.16) we obtain

$$\left.\begin{aligned} \Phi_* \xi_a &= A_a + \sum_{b=1}^{s} \lambda_a^b \xi'_b \\ \eta^a \circ \Phi_*^{-1} &= \sum_{b=1}^{s} \tilde{\lambda}_b^a \eta^{b'} \\ \Phi_* \circ \phi \circ \Phi_*^{-1} &= \phi' - \sum_{a=1}^{s} (\eta^a \circ \Phi_*^{-1}) \otimes JA_a, \end{aligned}\right\} \tag{4.2}$$

where $[\lambda_a^b]$ is an invertible matrix of differentiable functions on $\mathcal{U}'$, $[\tilde{\lambda}_a^b]$ is the inverse matrix of $[\lambda_a^b]$ and $A_a \in \Gamma(D')$.

Conversely, suppose $\Phi : M \to M'$ is a diffeomorphism satisfying (4.2) for two f-structures with complemented frames (ϕ, η^a, ξ_a) and $(\phi', \eta^{a'}, \xi'_a)$ associated to (D, J) and (D', J') respectively. Take $X \in \Gamma(D)$ and from the second equality in (4.2) we obtain

$$\eta^{b'}(\Phi_* X) = \sum_{a=1}^{s} \lambda_a^b \eta^a(X) = 0.$$

Hence $\Phi_* X \in \Gamma(D')$ for each $X \in \Gamma(D)$. Next, by using the third equality in (4.2), we get

$$\Phi_*(\phi X) = \phi'(\Phi_* X), \quad \text{for any} \quad X \in \Gamma(D).$$

But we know that the restrictions of ϕ and ϕ' to D and D' respectively are just J and J'. Consequently, we have $\Phi_* \circ J = J' \circ \Phi_*$, that is, Φ is a pseudo-conformal mapping. Thus we have the following important result (see Ishihara [2] for the case $s = 1$).

THEOREM 4.1. Let $\Phi : M \to M'$ be a diffeomorphism of two CR-manifolds M and M' with (D, J) and respectively (D', J') as CR-structures. Then Φ is a pseudo-conformal mapping if and only if there exist two f-structures with complemented frames (ϕ, η^a, ξ_a) and $(\phi', \eta^{a'}, \xi'_a)$ associated locally to (D, J) and (D', J') respectively such that (4.2) is satisfied.

Now, suppose $\Phi : M \to M$ is a pseudo-conformal mapping on the CR-manifold. Then we say that Φ is a pseudo-conformal transformation on M. Let X be a vector field on M. Then we say that X is a pseudo-conformal vector field if any local transformation Φ_t $(-\varepsilon < t < \varepsilon,\ \varepsilon > 0)$ of M spanned by X is a pseudo-conformal transformation. By using Theorem 4.1 we obtain

THEOREM 4.2. A vector field X is a pseudo-conformal vector field if and only if for any f-structure with complemented frames (ϕ, η^a, ξ_a) associated to the CR-structures (D, J) on a coordinate neighborhood $\mathcal{U}$ of M we have

$$\left.\begin{aligned} \mathcal{L}_X\phi &= -\sum_{b=1}^{s} \eta^b \otimes JB_b, \quad \mathcal{L}_X\eta^a = \sum_{b=1}^{s} \alpha^a_b\eta^b, \\ \mathcal{L}_X\xi_a &= -B_a - \sum_{b=1}^{s} \alpha^b_a\xi_b, \end{aligned}\right\} \tag{4.3}$$

where $[\alpha^a_b]$ are differentiable functions on $\mathcal{U}$, $B_b \in \Gamma(D)$ and $\mathcal{L}_X$ is the Lie derivative operator with respect to X.

On the other hand, we have

THEOREM 4.3. Let X be a pseudo-conformal vector field on a CR-manifold M and (D, J) be the CR-structure on M. If X belongs to D then $G^a(Y, X) = 0$, $a = 1,\dots,s$ for any vector field Y on M.

Proof. By (1.14) we see that $\phi Y \in \Gamma(D)$ for any $Y \in \Gamma(TM)$. Then (1.17) implies

$$G^a(Y, X) = \frac{1}{2}\{\phi Y(\eta^a X) - X(\eta^a(\phi Y)) - \\ - \eta^a([\phi Y, X])\} = -\frac{1}{2}\eta^a([\phi Y, X]). \tag{4.4}$$

Next, from the first equality in (4.3) we get

$$[X, \phi Y] - \phi([X, Y]) = - \sum_{b=1}^{s} \eta^b(Y) JB_b.$$

Hence $[X, \phi Y] \in \Gamma(D)$ and our assertion follows from (4.4).

We say that the CR-structure (D, J) is <u>non-degenerate</u> if at least one of the tensor fields G^a given by (1.17) is non-degenerate on D, that is, from $G^a(Y, X) = 0$ for any $Y \in \Gamma(D)$ and a certain $X \in \Gamma(D)$ it follows that $X = 0$. From (1.22) we see that the definition of a non-degenerate CR-structure does not depend on the associate f-structure with complemented frames. A CR-manifold whose CR-structure is non-degenerate is called a <u>non-degenerate CR-manifold</u>. Then by Theorem 4.3 we have

THEOREM 4.4. <u>Let X be a pseudo-conformal vector field on a non-degenerate CR-manifold M. If X belongs to D then X vanishes identically on M.</u>

If in particular, M is a real hypersurface of a complex space and the CR-structure on M is non-degenerate, Theorem 4.4 can be found in Tanaka [1] or Ishihara [2].

The remaining part of this section is devoted to the pseudo-conformal transformations on generic submanifolds.

Let $\tilde{M}$ be a complex manifold with almost complex structure $\tilde{J}$ and $\tilde{\Phi} : \tilde{M} \to \tilde{M}$ be a holomorphic transformation on $\tilde{M}$. Then we have $\tilde{\Phi}_* \circ \tilde{J} = \tilde{J} \circ \tilde{\Phi}_*$, where $\tilde{\Phi}_*$ denotes the differential mapping of $\tilde{\Phi}$. Next, we consider two generic submanifolds M and M' in $\tilde{M}$ such that $M' = \tilde{\Phi}(M)$. Denote by Φ the restriction of $\tilde{\Phi}$ to M and by (D, J) and (D', J') the CR-structures on M and M' respectively. We see that $\Phi_* X \in \Gamma(D')$ for each $X \in \Gamma(D)$ and $\Phi_* \circ J = J' \circ \Phi_*$, that is, Φ is a pseudo-conformal mapping. Therefore we have

THEOREM 4.5. <u>Any holomorphic transformation $\tilde{\Phi} : \tilde{M} \to \tilde{M}$ of the ambient complex manifold $\tilde{M}$ induces a pseudo-conformal mapping $\Phi : M \to M'$, where M and M' are generic submanifolds such that $M' = \tilde{\Phi}(M)$.</u>

Next from Theorems 2.1 and 4.1 we obtain

COROLLARY 4.1. <u>Let M and M' be two generic submanifolds of a complex manifold $\tilde{M}$ and $\Phi : M \to M'$ be a diffeomorphism. Then Φ is a pseudo-conformal transformation if and only if the equalities in (4.2) are satisfied.</u>

The following theorem can be obtained in the same fashion as Theorem 4.6 of Yano-Ishihara [3].

THEOREM 4.6. Let M and M' be two affine anti-holomorphic submanifolds analytically immersed in a complex manifold $\tilde{M}$ and assume $\Phi : M \to M'$ is an analytic homeomorphism. Then Φ is pseudo-conformal if and only if, for any point x belonging to M there exist neighborhoods $\tilde{U}$ and $\tilde{U}'$ of x and $\Phi(x)$ in $\tilde{M}$ respectively and a holomorphic homeomorphism $\tilde{\Phi} : \tilde{U} \to \tilde{U}'$ such that Φ is the restriction of $\tilde{\Phi}$ to $\tilde{U} \cap M$.

As a final remark we note that CR-manifolds of dimension 2n + 1 have been intensively studied by means of associated almost contact structures (see Ishihara [2], Yano-Ishihara [3], Sakamoto-Takemura [1], [2], Takemura [1]). The author believes that more results might be obtained on the geometry of CR manifolds of dimension 2n + s, via associate f-structures with complemented frames.

Chapter VII

CR-STRUCTURES AND RELATIVITY

The main purpose of this chapter is to show that CR-structures on real hypersurfaces of a complex manifold have an interesting application to relativity. It is the merit of Roger Penrose to discover a correspondence between points of a Minkowski space and projective lines of a certain real hypersurface in a complex projective space (see (2.4)).

§1. Geometrical Structures of Space-Time

In this section we summarize briefly the geometry of both the Minkowski space and the Einstein space, which in fact are the most important space-time models.

A Minkowski space is R^4 endowed with a flat pseudo-Riemannian metric g of Lorentzian signature (1, 3), (see §2 of Chapter I). We denote by M the Minkowski space defined above and by T_pM the tangent space to M at p. Then a tangent vector $X \in T_pM$ is said to be <u>null</u> if and only if $\|X\| = 0$, where $\|\cdot\|$ is the norm defined by g. The set of null vectors in T_pM is called the <u>null cone</u> or <u>light cone</u> at p and it is denoted by C_p.

In the case of Minkowski geometry it is convenient to work with null vectors as basis vectors. Suppose $\{X_0, X_1, X_2, X_3\}$ is a basis in T_pM. Then any $X \in T_pM$ is given by x^iX_i and $x = (x^0, x^1, x^2, x^3)$ gives a system of coordinates on M. We can choose the basis $\{X_i\}$ such that

$$g(X, Y) = x^0y^0 - x^1y^1 - x^2y^2 - x^3y^3$$

and denote $g(X, Y)$ by $g(x, y)$ and $\|x\| = \{g(x, x)\}^{1/2}$.

Now we construct a set of coordinates in $M \otimes_R C$ by the following equalities

$$\begin{cases} u = \frac{1}{\sqrt{2}}(x^0 + x^1) \\ \bar{\xi} = \frac{1}{\sqrt{2}}(x^2 - ix^3) \end{cases} \quad \begin{cases} \xi = \frac{1}{\sqrt{2}}(x^2 + ix^3) \\ v = \frac{1}{\sqrt{2}}(x^0 - x^1), \end{cases}$$

where $i = \sqrt{-1}$. Then we obtain $\|u\| = \|v\| = \|\xi\| = \|\bar{\xi}\| = 0$, where we considered the extension by complex linearity of g to $M \otimes_R C$. The complex matrix

$$A = \begin{bmatrix} u & \xi \\ \bar{\xi} & v \end{bmatrix}$$

has the determinant $\det(A) = uv - \xi\bar{\xi} = \frac{1}{2}\|x\|^2$ and it is a Hermitian matrix. On the other hand, each (2×2)-Hermitian matrix is of this form with some $(u, v, \xi, \bar{\xi})$. As we shall see later, it is convenient to identify the space H(2) of (2×2)-Hermitian matrices endowed with the determinant as norm with the Minkowski space in which we have choosen an origin and a basis for the tangent space at the origin.

Of course, both of the pseudometrics on M and $M \otimes_R C$ induce a distance in such spaces. We denote by d the <u>squared distance</u> function and then we have

$$d(x, x') = (x^0 - x^{0'})^2 - (x^1 - x^{1'})^2 - (x^2 - x^{2'}) - $$
$$- (x^3 - x^{3'})^2 = \det\left\{\begin{bmatrix} u - u' & \xi - \xi' \\ \bar{\xi} - \bar{\xi}' & v - v' \end{bmatrix}\right\}.$$

If we have $d(x, x') > 0$, $d(x, x') = 0$ or $d(x, x') < 0$ we say that the separation between x and x' is <u>timelike</u>, <u>null</u> or <u>spacelike</u> respectively.

The group of isometries of M is denoted by P and it is the 10-parameter <u>Poincaré group</u> and the <u>Lorentz group</u> at $p \in M$ is the subgroup L_p of P which leaves p fixed.

There exist four other alternatives of space-time structures: Aristotelean space-time, Galilean space-time, Newtonian space-time and Einsteinian space-time. For the geometrical structures of the first three space-time structures consult Penrose [2]. We give here only the geometrical structure of the Einsteinian space-time, which in fact can be regarded as the best model for gravity. As is well known, the mathematics of Einstein spaces is of great interest for physics.

An Einsteinian space-time is a 4-dimensional real manifold E endowed with a pseudometric g of signature (1, 3), but in this case g is generally non flat. The inertial motions in E are given by geodesics of the unique torsion-free connection ∇ induced by g. But this connection has a curvature R whose physical interpretation is obtained from the Jacobi equations. In order to provide a physical theory of gravity, Einstein had to postulate the so called <u>field equations</u> given by

$$S - \frac{1}{2}\rho g + \Lambda g = -\chi T,$$

where S is the Ricci tensor of g, ρ is the scalar curvature, χ is the Einstein's constant, Λ is the cosmological constant and T is the energy-momentum tensor. In fact we have here a system of ten partial differential equations of second order in which the ten components of the pseudometric g are the unknown functions. Einstein's model (whose particular case is Minkowski's model) is still the best of all space-time models known up to date.

§2. The Twistor Space and Penrose Correspondence

Let T be a 4-dimensional complex vector space and let (z^0, z^1, z^2, z^3) be the coordinates of an element z of T. Suppose on T there is given the Hermitian form Φ of signature (2, 2) defined by

$$\Phi(z) = z^0\bar{z}^2 + z^1\bar{z}^3 + z^2\bar{z}^0 + z^3\bar{z}^1. \tag{2.1}$$

The space T endowed with the Hermitian form Φ is called the <u>twistor space</u> (see Penrose [1], [2] and Wells [1]). Two points $z \in T$ and $x \in M$ (where M is a Minkowski space) are said to be incident whenever we have

$$(z^2, z^3) = \frac{1}{i\sqrt{2}}(z^0, z^1)\begin{bmatrix} u & \xi \\ \bar{\xi} & v \end{bmatrix}. \tag{2.2}$$

We see that (2.2) can only hold if $\Phi(z) = 0$.

Now, we denote by PT the complex 3-dimensional projective space associated with T. In this complex manifold we define the open complex submanifolds

$$PT_+ = \{Z \in PT;\ \Phi(Z) > 0\} \text{ and}$$

$$PT_- = \{Z \in PT,\ \Phi(Z) < 0\}$$

and a real 5-dimensional manifold

$$PT_0 = \{Z \in PT;\ \Phi(Z) = 0\}.$$

Of course, PT_0 is the common boundary of PT_+ and PT_- and thus it inherites a natural CR-structure as a real hypersurface of a complex manifold (see §1 of Chapter II). In this way PT_0 contains the points of PT for which (2.2) holds for a certain x, but not only them. As we can easily see, the points of projective line I, given by $Z^0 = Z^1 = 0$ admit no solution for x in (2.2).

For each $Z \in PT_0 - I$ we can solve (2.2) for a certain point $x \in M$. Moreover, the general solution is given by

$$\begin{bmatrix} u & \xi \\ \bar{\xi} & v \end{bmatrix} + k\begin{bmatrix} Z^1\bar{Z}^1 & -Z^1\bar{Z}^0 \\ -Z^0\bar{Z}^1 & Z^0\bar{Z}^0 \end{bmatrix} \tag{2.3}$$

where

$$x = \begin{bmatrix} u & \xi \\ \bar{\xi} & v \end{bmatrix}$$

is a solution and k is an arbitrary real number. All of these solutions are null vectors in M. Therefore, the points x incident with the given Z constitute a null geodesic (straight line) in M. Thus we can consider $PT_0 - I$ as the space of null geodesics in M.

On the other hand, for each $x \in M$ fixed in (2.2) we obtain a complex 2-dimensional linear subspace of T, that is, a complex projective line in PT. Moreover, this line lies in $PT_0 - I$ and each line lying entirely in $PT_0 - I$ is obtained from points in M. Hence we have the correspondence

$$\begin{gathered}\{\text{points of Minkowski space } M\} \\ \Downarrow \\ \{\text{projective lines lying entirely in } PT_0 - I\}.\end{gathered} \tag{2.4}$$

This correspondence will be called the Penrose correspondence. It is an interesting means of passing from the geometry of a Minkowski space to the geometry of a CR-manifold. By the Penrose correspondence the family of light rays through a fixed point x of the Minkowski space M represents the family of points of the corresponding complex projective line L_x in $PT_0 - I$. Taking account of the topology of L_x we conclude that the topology of an observer's field of vision is just S^2. More important is the holomorphic structure of L_x which shows up in the fields of vision of several observers passing through the same point x and which are conformally related to one another.

§3. Physical Interpretations of CR-Structures

As we have seen in the previous section, by the Penrose correspondence there appears the physical interpretation of the differential geometry determined by the CR-structure on the real 5-dimensional real hypersurface PT_0 in PT.

First, let us see what we can say about the almost complex distribution (D, J) on PT_0 (for the notations see §1 of Chapter VI). For any point Z of PT_0 we have a complex 2-dimensional holomorphic subspace D_Z on which the almost complex structure J is induced by that of PT. The point Z is interpreted as a light ray in the Minkowski space M. Now, suppose $x^0 = 0$, that is, time takes a specific value. Thus the geometrical picture becomes a Euclidean space E of real dimension 3. Then the "photon" described by Z is represented by a unit vector z at some point q of E. Hence we can imagine the tangent space to PT_0 as small displacements of q and z. If a displacement is such that q is moved in a direction which is orthogonal to the direction of z we get an element of D_Z. The subspace D_Z divides the tangent space of PT_0 at Z into two remaining pieces representing light rays that are slightly ahead or lag slightly behind the original ray Z. All of these properties are independent of time x^0, that is, the above interpretation of D_Z is independent of the Euclidean hyperplane E.

Next, for the physical interpretation of J we imagine a 2-plane element π at q, which is orthogonal to the direction of the unit vector z. We are interested in studying displacements of q and z such that q is moved within π, since these correspond to real vectors in D_Z. Such displacements are represented by pairs of vectors (r, v) in π, where r gives the displacement of q and v measures the change in z. Now keep the light ray Z and the neighbouring ray to which it is displaced fixed and vary the time x^0. Then the functions $r(x^0)$ and $v(x^0)$ are related by

$$\frac{dr}{dx^0} = v \quad \text{and} \quad \frac{dv}{dx^0} = 0 \tag{3.1}$$

The action of the operator J means to rotate both r and v through a right angle in the plane π, in a left-handed sense about the direction of z. Moreover, (3.1) are invariant with respect to this action.

In this way, we obtained for the CR-structure of PT_0 an interpretation in terms of physical space - time geometry. On the other hand, PT_0 is just $S^2 \times S^3$ (see Wells [1]) and thus we know completely the geometry of PT_0. With this example in mind we can think of the following problem of differential geometry. Let N be a real 5-dimensional CR-manifold. Then under what conditions does N become a Penrose hypersurface, that is, when can N be immersed in CP^3 as a hypersurface which is pseudo-conformal (see §4 of Chapter VI) with PT_0?

CR-structures can also be discussed in the general context of a Einsteinian space-time manifold. But in this case we have to remark that the induced CR-structures might be locally distinct from one another.

As we have seen in this chapter, the CR-manifolds which appear in relativity by means of the Penrose correspondence are in fact real hypersurfaces of a certain complex manifold. However, counterexamples given by Nirenberg [1] show that not all CR-manifolds can be realized as a real hypersurface in a complex manifold. The CR-structures of such CR-manifolds are called nonrealizable CR-structures and they are now intensively studied by many people.

REFERENCES

Andreotti A. and Hill C.D.:

1. 'Complex characteristic coordinates and tangential Cauchy-Riemann equations', Ann. Scuola Norm. Sup. Pisa Sci. Fis. Nat. 26 (1972), 299-324.

Arca G. and Roşca R.:

1. 'Contact CR-submanifolds of a Sasakian manifold admitting a contact concircular vector pairing', Tensor N.S. 40 (1983), 280-284.

Barros M. and Urbano F.:

1. 'CR-submanifolds of generalized complex space forms', An. St. Univ. Al. I. Cuza Iaşi 25 (1979), 355-365
2. 'Submanifolds of complex Euclidean space which admit a holomorphic distribution', Quart. J. Math. Oxford 34 (1983), 141-143.
3. 'Topology of quaternion CR-submanifolds', Bollettino U.M.I. 6 (1983), 103-110.

Barros M., Chen B.Y., and Urbano F.:

1. 'Quaternion CR-submanifolds of quaternion manifolds', Kodai Math. J. 4 (1981), 399-418.

Bejancu A.:

1. 'CR-submanifolds of a Kaehler manifold I', Proc. Amer. Math. Soc. 69 (1978), 134-142.
2. 'On the integrability conditions on a CR-submanifold', An. St. Univ. Al. I. Cuza Iaşi 24 (1978), 21-24.
3. 'Une classe de sous-variétés d'une variété kählerienne', C. R. Acad. Sc. Paris 286 (1978), (Série A), 597-599.
4. 'CR-submanifolds of a Kaehler manifold II', Trans. Amer. Math. Soc. 250 (1979), 333-345.

5. 'On the geometry of leaves on a CR-submanifold', An. St. Univ. Al. I. Cuza Iaşi 25 (1979), 393-398.

6. 'Real hypersurfaces of a complex projective space', Rendiconti di Mat. 12 (1979), Serie VI, 439-445.

7. 'Normal CR-submanifolds of a Kaehler manifold', An. St. Univ. Al. I. Cuza Iaşi 26 (1980), 123-132.

8. 'Umbilical CR-submanifolds of a Kaehler manifold', Rendiconti di Mat. 13 (1980), 431-446.

9. 'On a class of mixed totally geodesic CR-submanifolds', An. Univ. Timişoara 28 (1980), 11-23.

10. 'F-connections on a CR-submanifold', Bul. Inst. Politehnic Iasi 27 (1981), 33-40.

11. 'Hypersurfaces of quaternion space forms', An. St. Univ. Al. I. Cuza Iaşi 27 (1981), 291-296.

12. 'On semi-invariant submanifolds of an almost contact metric manifold', An. St. Univ. Al. I. Cuza Iaşi, Supliment tom 27 (1981), 17-21.

13. 'Umbilical semi-invariant submanifolds of a Sasakian manifold', Tensor N.S. 37 (1982), 203-213.

14. 'Sasakian anti-holomorphic submanifolds of a Kaehler manifold', Glasnik Matematicki 17 (1982), 115-130.

15. 'Anti-holomorphic submanifolds of almost Hermitian manifolds', Math. Rep. Toyama Univ. 6 (1983), 179-196.

16. 'A theorem of classification for semi-invariant submanifolds of a Sasakian space form', An. St. Univ. Al. I. Cuza Iaşi 29 (1983), 59-64.

17. 'Hypersurfaces of quaternion manifolds', Revue Roumaine de Math. Pures et Appl. 28 (1983), 567-576.

18. 'Pinching theorems for sectional curvature of a CR-submanifold', Rendiconti di Mat. 3 (1983), 65-71.

19. 'On contact umbilical submanifolds of Sasakian space forms', An. St. Univ. Al. I. Cuza Iaşi 30 (1984), 89-94.

20. 'Hypersurfaces of manifolds with a Sasakian 3-structure', Demonstratio Math. 17 (1984), 197-209.

21. 'Semi-invariant submanifolds of locally product Riemannian manifolds', An. Univ. Timişoara 22 (1984), 3-11.

22. 'Anti-quaternion submanifolds of quaternion manifolds', Lucrările Conf. Geom. Top., P. Neamt, (1984), 141-144.

23. 'CR-structures and pseudo-conformal mappings', Seminarul Geom. Top. nr. 78, Univ. Timişoara, 1984.

24. 'QR-submanifolds of quaternion manifolds', to appear.

25. 'Generic submanifolds of manifolds with a Sasakian 3-structure', Math. Rep. Toyama Univ. 8 (1985), 75-101.

Bejancu A., Kon M., and Yano K.:

1. 'CR-submanifolds of a complex space form', J. Differential Geometry 16 (1981), 137-145.

Bejancu A. and Papaghiuc N.:

1. 'Semi-invariant submanifolds of a Sasakian manifold', An. St. Univ. Al. I. Cuza Iaşi 27 (1981), 163-170.
2. 'Semi-invariant submanifolds of a Sasakian space form', Colloquium Math. 48 (1984), f.2 , 229-240.
3. 'Almost semi-invariant submanifolds of a Sasakian manifold', Bull. Math. Soc. Sci. R.S. Roumanie 28 (1984), 321-338.

Bejancu A. and Smaranda D.:

1. 'Semi-invariant submanifolds of a co-Kähler manifold', An. St. Univ. Al. I. Cuza Iaşi 29 (1983), 27-32.

Bishop R.L. and Goldberg S.I.:

1. 'Some implications of the generalized Gauss-Bonnet theorem', Trans. Amer. Math. Soc. 112 (1964), 508-535.

Blair D.E.:

1. 'Geometry of manifolds with structural group $\mathcal{U}(n) \times \mathcal{O}(s)$', J. Differential Geometry 4 (1970), 155-167.
2. 'On a generalization of the Hopf fibration', An. St. Univ. Al. I. Cuza Iaşi 17 (1971), 171-177.
3. 'Contact manifolds in Riemannian geometry', Lecture Notes in Math. 509, Springer Verlag, Berlin, 1976.
4. 'Three lectures on complex differential geometry', Bull. Soc. Mat. de Belgique 35 (1983), 25-38.

Blair D.E. and Chen B.Y.:

1. 'On CR-submanifolds of Hermitian manifolds', Israel J.

Math. 34 (1979), 353-363.

Blair D.E., Ludden G.D., and Yano K.:
1. 'Differential geometric structures on principal toroidal bundles', Trans. Amer. Math. Soc. 181 (1973), 175-184.

Buchner K. and Roşca R.:
1. 'Sasakian manifolds having the contact quasi-concurrent property', Rendiconti del Circolo Mat. di Palermo 32 (1983), 388-397.

Calabi E.:
1. 'Metric Riemannian surfaces', Annals of Math. Studies No. 30, Princeton University Press, Princeton.
2. 'Isometric imbedding of complex manifolds', Annals of Math. 58 (1953), 1-23.

Calapso M.T. and Roşca R.:
1. 'Sous variétés génériques pseudo-ombilicales de contact d'une variété sasakienne', Rendiconti del Circolo Mat. di Palermo 32 (1983), 69-75.

Cartan E.:
1. 'Sur la géométrie pseudo-conforme des hypersurfaces de deux variables complexes, I, II', Ann. Math. Pura Appl. 11 (1932), 17-90; Ann. Scuola Norm. Sup. Pisa 1 (1932), 333-354.

Chen B.Y.:
1. Geometry of Submanifolds, M. Dekker Inc., New York, 1973.
2. 'Totally umbilical submanifolds of quaternion space forms', J. Austral. Math. Soc. 26 (1978), 154-162.
3. 'Totally umbilical submanifolds', Soochow Journal of Math. 5 (1979), 9-37.
4. 'Totally umbilical submanifolds of Kaehler manifolds', Archiv der Mathematik 36 (1981), 83-91.
5. 'CR-submanifolds of a Kaehler manifold I', J. Differential Geometry 16 (1981), 305-323.
6. 'CR-submanifolds of a Kaehler manifold II', J. Differential Geometry 16 (1981), 493-509.

7. 'Cohomology of CR-submanifolds', An. Faculté des Sciences Toulouse 3 (1981), 167-172.

8. Geometry of Submanifolds and Its Applications, Science University of Tokyo, 1981.

9. 'Differential geometry of real submanifolds in a Kaehler manifold', Monat. für Math. 91 (1981), 257-274.

10. 'Some non-integrability theorems of holomorphic distributions', to appear.

Chen B.Y., Ludden G.D., and Montiel S.:

1. 'Real submanifolds of a Kaehler manifold', Algebras, Groups and Geometries 1 (1984), 176-212.

Chen B.Y. and Lue H.S.:

1. 'On normal connections of Kaehler submanifolds', J. Math. Soc. Japan 27 (1975), 550-556.

Chen B.Y. and Ogiue K.:

1. 'On the scalar curvature and sectional curvature of a Kaehler submanifold', Proc. Amer. Math. Soc. 41 (1973), 247-250.

2. 'Some extrinsic results for Kaehler submanifolds', Tamkang J. Math. 4 (1973), 207-213.

3. 'On totally real submanifolds', Trans Amer. Math. Soc. 193 (1974), 257-266.

4. 'Two theorems on Kaehler manifolds', Michigan Math. J. 21 (1974), 225-229.

Chen B.Y. and Okumura M.:

1. 'Scalar curvature, inequality and submanifolds', Proc. Amer. Math. Soc. 38 (1973), 605-608.

Chern S.S. and Moser J.K.:

1. 'Real hypersurfaces in complex manifolds', Acta Math. 133 (1974), 219-271.

Cihodariu Gh. and Smaranda D.:

1. 'Semi-invariant submanifolds of a manifold with Sasakian (f, g, u, v)-structure', Lucrările Coloc. Nat. Geom. Top., Busteni (1981), 54-64.

Fukami T. and Ishihara S.:
1. 'Almost Hermitian structure on S^6', Tôhoku Math. J. 7 (1955), 151-156.

Gheorghiev Gh. and Oproiu V.:
1. 'Finite and infinite dimensional differentiable manifolds I, II', Ed. Acad. R.S.R., București, 1976, 1979, (in Romanian).

Goldberg S.I. and Yano K.:
1. 'On normal globally framed f-manifolds', Tôhoku Math. J. 22 (1970), 362-370.
2. 'Globally framed f-manifolds', Illinois J. of Math., 15 (1971), 456-474.

Goldberg V.V. and Roșca R.:
1. 'Mixed isotropic submanifolds and isotropic cosymplectic structures', Soochow J. of Math. 9 (1983), 71-84.

Gray A.:
1. 'Kaehler submanifolds of homogeneous almost Hermitian manifolds', Tôhoku Math. J. 21 (1969), 190-194.
2. 'Six dimensional almost complex manifolds defined by means of three-fold vector cross products', Tôhoku Math. J. 21 (1969), 614-620.
3. 'Almost complex submanifolds of the six sphere', Proc. Amer. Math. Soc. 20 (1969), 277-279.

Greenfield S.:
1. 'Cauchy-Riemann equations in several variables', An. della Scuola Norm. Sup. Pisa 22 (1968), 275-314.

Hsu C.J.:
1. 'On CR-submanifolds of Sasakian manifolds I', Math. Research Center Reports, Symposium Summer 1983, 117-140.
2. 'On some properties of CR-submanifolds of Kaehler manifolds', Chinese J. of Math. 22 (1984), 7-27.
3. 'Two theorems on CR-submanifolds of Kaehler manifolds', Bull. of Inst. Math. Acad. Sinica 12 (1984), 95-99.

Ianuș, S.:
1. 'Sulle varietà di Cauchy-Riemann', Rend. dell'Accademia

di Science Fisiche Mat. Napoli 33 (1972), 191-195.

Ianuş S. and Mihai I.:
1. 'Semi-invariant submanifolds of an almost paracontact manifold', Tensor N.S. 39 (1982), 195-199.

Ianuş S., Mihai I., and Matsumoto K.:
1. 'Almost semi-invariant submanifolds of some almost paracontact Riemannian manifolds', Bull. of the Yamagata Univ. 11 (1985), 121-128.

Ishihara S.:
1. 'Quaternion Kaehlerian manifolds', J. Differential Geometry 9 (1974), 483-500.
2. 'Distributions with complex structure', Kodai Math. J. 1 (1978), 264-276.

Ki U.H. and Jin D.H.:
1. 'Generic submanifolds with parallel Ricci curvature of $S^{2n+1}(1)$', J. Korean Math. Soc. 19 (1982), 55-60.

Ki U.H., Pak J.S., and Kim Y.H.:
1. 'CR-structures of submanifolds immersed in complex space forms', Kyungpook Math. J. 23 (1983), 155-168.

Ki U.H. and Pak J.S.:
1. 'Generic submanifolds of an even-dimensional Euclidean space', J. Differential Geometry 16 (1983), 293-303.

Kim I.B.:
1. 'Submanifolds of Kaehlerian manifolds and metric compound structures', Hiroshima Math. J. 13 (1983), 401-443.

Kobayashi S. and Nomizu K.
1. Foundation of Differential Geometry, I, II, Interscience, New York, 1963, 1969.

Kobayashi M.:
1. 'CR-submanifolds of a Sasakian manifold', Tensor N.S. 35 (1981), 297-307.
2. 'CR-submanifolds of a Sasakian space form with flat normal connection', Tensor N.S. 36 (1982), 207-213.

3. 'Integrabilities of CR-submanifolds of a nearly Sasakian manifold', Tensor N.S. 36 (1982), 215-221.

4. 'Contact CR-products of Sasakian manifolds', Tensor N.S. 36 (1982), 281-287.

5. 'Symmetric twofold CR-submanifolds in a Euclidean space R^{4n}', to appear.

Kon M.:

1. 'On some complex submanifolds in Kaehler manifolds', Canadian J. Math. 26 (1974), 1442-1449.

2. 'Invariant submanifolds in Sasakian manifolds', Math. Ann. 219 (1976), 277-290.

Lawson H.B. Jr.:

1. 'Rigidity theorems in rank-1 symmetric spaces', J. Differential Geometry 4 (1970), 349-357.

Ludden G.D.:

1. 'Submanifolds of cosymplectic manifolds', J. Differential Geometry 4 (1970), 237-244.

Ludden G.D., Okumura M., and Yano K.:

1. 'Totally real submanifolds of complex manifolds', Atti della Acad. Naz. Lincei 58 (1975), 346-353.

Maeda Y.:

1. 'On real hypersurfaces of a complex projective space', J. Math. Soc. of Japan 28 (1976), 529-540.

Maeda S. and Sato N.:

1. 'On submanifolds all of whose geodesics are circles in a complex space form', Kodai Math. J. 6 (1983), 157-166.

Martinez A., Pérez J.D., and Santos F.G.:

1. 'On the normal connection of quaternion CR-submanifolds', Bull. of Inst. Math. Acad. Sinica 12 (1984), 237-247.

Martinez A. and Santos F.G.:

1. 'Generic submanifolds of quaternion Kaehlerian manifold', to appear.

Matsumoto K.:
1. 'On submanifolds of locally product Riemannian manifolds', TRU Mathematics 18-2 (1982), 145-157.
2. 'On contact CR-submanifolds of Sasakian manifolds', Internat. J. Math. Sci. 6 (1983), 313-326.
3. 'On CR-submanifolds of locally conformal Kähler manifolds', J. Korean Math. Soc. 21 (1984), 49-61.

Mihai I.:
1. 'CR-submanifolds of a framed f-manifold', Stud. Cerc. Mat. 35 (1982), 127-136.

Miron R.:
1. Differential Geometry, Ed. Did. Ped., Bucureşti, 1976, (in Romanian).

Moore J.D.:
1. 'Isometric immersions of Riemannian products', J. Differential Geometry 5 (1971), 159-168.

Naitza D.:
1. 'CR-sous-variétés de contact à forme simple verticale fermées d'une variété sasakienne', Rend. Sem. Fac. Sci. Univ. Cagliari 52 (1982), 31-34.

Nirenberg L.:
1. Lectures on Linear Partial Differential Equations, CBMS Regional Conf. Ser. in Math. No 17, Amer. Math. Soc., Providence R.I., 1973.

Ogiue K.:
1. 'Differential geometry of Kaehler submanifolds', Advances in Math. 13 (1974), 73-114.

Okumura M.:
1. 'Certain almost contact hypersurfaces in Kaehlerian manifolds of constant holomorphic sectional curvature', Tôkohu Math. J. 16 (1964), 270-284.
2. 'Submanifolds and a pinching problem on the second fundamental form', Trans. Amer. Math. Soc. 178 (1973), 285-291.
3. 'On some real hypersurfaces of a complex projective space', Trans. Amer. Math. Soc. 212 (1975), 355-364.

4. 'Submanifolds of real codimension p of complex projective space', Atti della Accad. Naz. Lincei 58 (1975), 544-555.
5. 'Compact real hypersurfaces of a complex projective space', J. Differential Geometry 12 (1977), 595-598.

O'Neill B.:
1. Semi-Riemannian Geometry, Academic Press, New York, 1983.

Oproiu V.:
1. Varietà di Cauchy-Riemann, Institute di Mat. dell Univ. di Napoli, Relatione No. 20, 1972.

Ornea L.:
1. 'Generic CR-submanifolds of S-manifolds', Stud. Cerc. Mat. 36 (1984), 435-443.

Pak, J.S.:
1. 'Real hypersurfaces in quaternion Kaehlerian manifolds with constant Q-sectional curvature', Kodai Math. Sem. Rep. 29 (1977), 22-61.
2. 'Anti-quaternion submanifolds of a quaternion projective space', Kyungpook Math. J. 18 (1981), 91-115.
3. 'Contact-three-CR-submanifolds', J. Korean Math. Soc. 19 (1982), 1-10).
4. 'CR- and contact CR-submanifolds', Research Review of Kyungpook Nat. Univ. 35 (1983), 225-230.
5. 'Quaternionic CR-submanifolds of quaternionic space form', to appear.

Pak J.S. and Kang T.H.:
1. 'Generic submanifolds of a quaternionic projective space with parallel mean curvature vector', J. Korean Math. Soc. 17 (1981), 175-192.

Papaghiuc N.:
1. 'Semi-invariant submanifolds in Kenmotsu manifolds', Rendiconti di Mat. 3 (1983), 607-622.
2. 'Semi-invariant products in Sasakian manifolds', An. St. Univ. Al. I. Cuza Iaşi 30 (1984), 69-78.

Papuc D.:
1. Differential Geometry, Ed. Did. Ped., Bucureşti (1982),

(in Romanian).

Penrose R.:
1. 'The complex geometry of the natural world', Proc. of Internat. Congres of Math., Helsinki (1978), 189-194.
2. 'Physical space-time and nonrealizable CR-structures', Proc. of Symposia in Pure Math. 39 (1983), 401-422.

Ros A.:
1. 'Spectral geometry of CR-minimal submanifolds in the complex projective space', Kodai Math. J. 6 (1983), 88-99.

Ros A. and Verstraelen L.:
1. 'Solution of a conjecture by K. Ogiue concerning the sectional curvature of Kaehler submanifolds', to appear.

Roşca. R.:
1. 'CR-hypersurfaces à champs normal covariant décomposable inclues dans une variété pseudo-riemannienne neutre', C.R. Acad. Sc. Paris, (Série A) 292 (1981), 287-290.
2. 'Sous variétés génériques d'une variété sasakienne à champ vectorial anti-invariant Φ-récurrent', Riv. Mat. Univ. Parma 9 (1983), 125-131.

Sakamoto K. and Takemura Y.:
1. 'On almost contact structures belonging to a CR-structure', Kodai Math. J. 3 (1980), 144-161.
2. 'Curvature invariants of CR-manifolds', Kodai Math. J. 4 (1980), 251-265.

Sasaki S.:
1. Almost Contact Manifolds, Lecture Notes, Tôhoku Univ., 1965.

Sasaki S. and Hatakeyama Y.:
1. 'On differentiable manifolds with certain structures which are closely related to almost contact structures II', Tôhoku Math. J. 13 (1961), 281-294.

Sato N.:
1. 'Certain anti-holomorphic submanifolds of almost Hermitian manifolds', Science Reports of Niigata Univ. 18 (1982), 1-9.

2. 'Certain CR-submanifolds of almost Hermitian manifolds', to appear.

3. 'On certain CR-submanifolds of 6-dimensional nearly Kaehlerian manifolds', to appear.

4. 'On anti-holomorphic and CR-submanifolds of a complex projective space', to appear.

Sekigawa K.:
1. 'Some CR-submanifolds in a 6-dimensional sphere', Tensor N.S. 41 (1984), 13-20.

Shimizu Y.:
1. 'On a construction of homogeneous CR-submanifolds in a complex projective space', Commentarii Math. Univ. Sancti Pauli 32 (1983), 203-207.

Simionescu S.:
1. 'CR-submanifolds of codimension two and (f, g, u, v, λ)-structures', to appear.

Smyth B.:
1. 'Differential geometry of complex hypersurfaces', Ann. of Math. 85 (1967), 247-266.

2. 'Homogeneous complex hypersurfaces', J. Math. Soc. Japan 20 (1968), 643-647.

Stong R.E.:
1. 'The rank of an f-structure', Kodai Math. Sem. Rep. 29 (1977), 207-209.

Takagi R.:
1. 'On homogeneous real hypersurfaces in a complex projective space', Osaka J. Math. 10 (1973), 495-506.

2. 'Real hypersurfaces in a complex projective space with constant principal curvatures, I, II', J. Math. Soc. Japan 27 (1975), 43-53, 507-516.

Takemura Y.:
1. 'On the invariant submanifold of a CR-manifold', Kodai Math. J. 5 (1982), 416-425.

Tanaka N.:
1. 'On the pseudo-conformal geometry of hypersurfaces of

the space of n complex variables', J. Math. Soc. Japan 14 (1962), 397-429.

2. 'On non-degenerate real hypersurfaces, graded Lie algebras and Cartan connections', Jap. J. Math., New Ser. 2 (1976), 131-190.

Tashiro Y.:
1. 'On contact structure of hypersurfaces in complex manifolds I, II', Tôhoku Math. J. 15 (1963), 62-78; 167-175.

Tashiro Y. and Kim I.B.:
1. On almost contact metric compound structure, Kodai Math. J. 5 (1982), 13-29.

Teleman C.:
1. Local and Global Differential Geometry, Ed. Tehnică, Bucuresti, 1974 (in Romanian).

Urbano F.:
1. CR-submanifolds. Totally real submanifolds of quaternion manifolds, Thesis Univ. Granada, 1979 (in Spanish).

Vaisman I.:
1. 'On locally conformal almost Kaehler manifolds', Israel J. Math. 24 (1976), 338-351.

Vanhecke L.:
1. 'On immersions with trivial normal connection in some almost Hermitian manifolds', Rendiconti di Mat. 10 (1977), 75-86.

Verheyen P.:
1. Generic Submanifolds of Sasakian Manifolds, Mededelingen uit het Wiskundig Instituut, Katholieke Univ. Leuven, No. 157, 1982.

Verstraelen L.:
1. Three Types of Results in Differential Geometry of Submanifolds, Mededelingen uit het Wiskundig Instituut, Katholieke Univ. Leuven, No. 154, 1982.

Walker A.G.:
1. 'Connections for parallel distributions in the large I, II', Quart. Math. Oxford 6 (1955), 301-308; 9 (1958), 221-231.

Wells R.O. Jr.:

1. 'Complex manifolds and mathematical physics', Bull. Amer. Math. Soc. (N.S.) 1 (1979), 296-336.
2. Differential Analysis on Complex Manifolds, New York, Berlin, Springer, 1980.
3. 'The Cauchy-Riemann equations and differential geometry', Proc. of Symposia in Pure Math. 39 (1983), 423-435.

Willmore T.J.:

1. 'Parallel distributions on manifolds', Proc. London Math. Soc. 6 (1956), 191-204.
2. Total Curvature in Riemannian Geometry, Ellis Horwood Limited, 1982.

Yano K.:

1. 'On a structure defined by a tensor field of type (1, 1) satisfying $f^3 + f = 0$, Tensor N.S. 14 (1963), 99-109.
2. Differential Geometry on Complex and Almost Complex Spaces, Pergamon, New York, 1965.

Yano K. and Ishihara S.:

1. 'Almost contact structures induced on hypersurfaces in complex and almost complex spaces', Kodai Math. Sem. Reports 17 (1965), 222-249.
2. 'Submanifolds with parallel mean curvature', J. Differential Geometry 6 (1971), 95-118.
3. 'Real hypersurfaces of a complex manifold and distributions with complex structure', Kodai Math. J. 1 (1978), 289-303.

Yano K. and Kon M.:

1. Anti-invariant Submanifolds, M. Dekker Inc., New York, 1976.
2. 'CR-sous-variétés d'un espace projectif complexe', C.R. Acad. Sc. Paris 288 (1979), 515-517.
3. 'Generic submanifolds', Ann. di Mat. Pura Appl. 123 (1980), 59-92.
4. 'Generic submanifolds of Sasakian manifolds', Kodai Math. J. 3 (1980), 163-196.
5. 'Differential geometry of CR-submanifolds', Geometriae

Dedicata 10 (1981), 369-391,

6. 'CR-submanifolds of a complex projective space', J. Differential Geometry 16 (1981), 431-444.

7. 'Contact CR-submanifolds', Kodai Math. J. 5 (1982), 238-252.

8. CR-Submanifolds of Kaehlerian and Sasakian Manifolds, Birkhäuser, Boston, 1983.

Yoo I.Y.:

1. 'Einstein CR-submanifolds of a flat Kaehlerian manifold', Honam Math. J. 6 (1984), 117-122.

AUTHOR INDEX

SUBJECT INDEX

Mathematics and Its Applications

Managing Editor:

Mathematics and Its Applications is devoted to interrelations such as:
- a central concept which plays an important role in several different specialized areas of mathematics and/or science;
- new applications of the results and ideas from one area of scientific endeavour into another;
- the influence which the results, problems and concepts of one field of enquiry have, and have had, on the development of another.

Present-day (applied) mathematics has some noteworthy features compared with even a few decennia ago:
- Rapidly increasing mathematization of a large number of areas such as, e.g., physical chemistry, electrical engineering, and geology.
- The enormous power of mathematical modeling which, when combined with powerful computers, enables us to carry out experiments which otherwise "in reality" would be too dangerous, too expensive, take too long, and be simply impossible.
- The era of increasing specialization seems to be over. The time has come to start using all the powerful tools that have been forged.

All in all it looks like the next decennia will witness an unequalled flowering in the mathematical sciences.

Based on these facts and this philosophy, the *Mathematics and Its Applications* book series is being developed quite vigorously. Given its success we plan to continue along the same lines often by means of specially commissioned volumes covering new areas of interaction likely to generate new kinds of problems, or covering a more established topic together with all its interrelations with others.

MAIN SERIES

W. KUYK. **Complementarity in Mathematics.** A First Introduction to the Foundations of Mathematics and Its History.
P. H. SELLERS. **Combinatorial Complexes.** A Mathematical Theory of Algorithms.
J. CHAILLOU. **Hyperbolic Differential Polynomials and Their Singular Perturbations.**
S. FUČÍK. **Solvability of Nonlinear Equations and Boundary Value Problems.**
W. L. MIRANKER. **Numerical Methods for Stiff Equations and Singular Perturbation Problems.**

P. M. COHN. **Universal Algebra.**

V. I. ISTRĂŢESCU. **Fixed Point Theory.** An Introduction.

N. E. HURT. **Geometric Quantization in Action.** Applications of Harmonic Analysis in Quantum Statistical Mechanics and Quantum Field Theory.

P. M. ALBERTI and A. UHLMANN. **Stochasticity and Partial Order.** Doubly Stochastic Maps and Unitary Mixing.

L. LANGOUCHE, D. ROEKAERTS, and E. TIRAPEGUI. **Functional Integration and Semiclassical Expansions.**

C. P. BRUTER, A. ARAGNOL, and A. LICHNEROWICZ (eds). **Bifurcation Theory, Mechanics and Physics.** Mathematical Developments and Applications.

J. ACZÉL (ed.). **Functional Equations: History, Applications and Theory.**

A. V. ARKHANGEL'SKII and V. I. PONOMAREV. **Fundamentals of General Topology.** Problems and Exercises.

P. E. T. JØRGENSEN and R. T. MOORE. **Operator Commutation Relations.**

R. BELLMAN and G. ADOMIAN. **Partial Differential Equations.** New Methods for Their Treatment and Solution.

N. K. BOSE (ed.). **Multidimensional Systems Theory.** Progress, Directions and Open Problems in Multidimensional Systems.

R. BELLMAN and R. VASUDEVAN. **Wave Propagation – an Invariant Imbedding Approach.**

R. A. ASKEY, T. H. KOORNWINDER, and W. SCHEMPP (eds.). **Special Functions: Group Theoretical Aspects and Applications**

L. ARNOLD and P. KOTELENEZ (eds). **Stochastic Space-Time Models and Limit Theorems.**

Y. BEN-HAIM. **The Assay of Spatially Random Material.**

Y. CHERRUAULT. **Mathematical Modelling in Biomedicine: Optimal Control of Biomedical Systems.**

R. E. BELLMAN and R. S. ROTH. **Methods in Approximation.** Techniques for Mathematical Modelling.

V. KOMKOV. **Variational Principles of Continuum Mechanics with Engineering Applications.** Volume 1: Critical Points Theory.

C. CUVELIER, A. SEGAL, and A. A. VAN STEENHOVEN. **Finite-Element Methods and Navier-Stokes Equations.**

J. D. LOUCK and N. METROPOLIS. **Symbolic Dynamics of Trapezoidal Maps.**

A. G. RAMM. **Scattering by Obstacles.**

EAST EUROPEAN SERIES

F.-H. VASILESCU. **Analytic Functional Calculus and Spectral Decompositions.**

W. KECS. **The Convolution Product and Some Applications.**

K. REKTORYS. **The Method of Discretization in Time** and Partial Differential Equations.

J. MŎCKOŘ. **Groups of Divisibility.**

L. IXARU. **Numerical Methods for Differential Equations and Applications.**

H. WALTHER. **Ten Applications of Graph Theory.**

A. WAWRZYŃCZYK. **Group Representations and Special Functions.**

D. S. MITRINOVIĆ and J. KEČKIC. **The Cauchy Method of Residues.** Theory and Applications.

V. BARBU and T. PRECUPANU. **Convexity and Optimization in Banach Spaces.**

I. BUCUR. **Selected Topics in Algebra** and its Interrelations with Logic, Number Theory and Algebraic Geometry.

I. M. STANCU-MINASIAN. **Stochastic Programming with Multiple Objective Functions.**

J. SZÉP and F. FORGÓ. **Introduction to the Theory of Games.**

L. BERAN. **Orthomodular Lattices.** Algebraic Approach.

A. MARCINIAK. **Numerical Solutions of the N-Body Problem.**

S. ROLEWICZ. **Metric Linear Spaces.**

A. PAZMÁN. **Foundations of Optimum Experimental Design.**

K. WAGNER and G. WECHSUNG. **Computational Complexity.**

SOVIET SERIES

A. V. SKOROHOD. **Random Linear Operators.**

B. S. RAZUMIKHIN. **Physical Models and Equilibrium Methods in Programming and Economics.**

N. H. IBRAGIMOV. **Transformation Groups Applied to Mathematical Physics.**

A. S. DAVYDOV. **Solitons in Molecular Systems.**